Metaphysics &
Medicine

Metaphysics & Medicine

Restoring Freedom of Thought to the Art and Science of Healing

Larry Malerba, D.O.

𝓜

Maverick Press
New York

Metaphysics & Medicine
Restoring Freedom of Thought to the Art and Science of Healing

First Edition November, 2014
Maverick Press, Altamont, New York

Cover Art and Design © The Metaphysician's Hand by Joel Malerba, inspired by the ancient alchemical Philosopher's Hand

Typesetting by FormattingExperts.com

For more information, contact Maverick Press:
maverickpressbooks@gmail.com

Medical Disclaimer: All content in this book is intended for educational purposes only and should not be construed as medical advice or any other kind of professional advice for anyone reading these materials. This information is not intended as a substitute for consultation with a qualified health care professional. All readers are encouraged to seek appropriate counsel in the event of a health problem.

ISBN-13: 978-1503055797

Printed in the USA

Contents

Foreword by Don Salmon 1

Introduction 7

Part I: The Many *-isms* of Medicine

CHAPTER 1: Medical Presuppositions 23

CHAPTER 2: Medical Materialism 33

CHAPTER 3: Medical Reductionism 45

CHAPTER 4: Medical Mechanism 55

CHAPTER 5: Medical Rationalism 71

CHAPTER 6: Medical Objectivism 89

CHAPTER 7: Medical Empiricism 119

CHAPTER 8: Medical Conformism 137

CHAPTER 9: Medical Dualism 153

Part II: Medical Claims to Knowledge

CHAPTER 10: Information, Knowledge & Abstraction 169

CHAPTER 11: Description, Explanation & Speculation 179

CHAPTER 12: Trials and Tribulations 197

CHAPTER 13: Medical Reality 217

CHAPTER 14: Scientism, Skepticism & Medical Fundamentalism 231

Part III: Authentic Medical Science

CHAPTER 15: Toward a New Set of Assumptions 271

CHAPTER 16: A Clinical Case Study 319

CHAPTER 17: The Rebirth of Medical Philosophy 331

Glossary Of Terms 345

Epigraph Credits & Permissions 353

Notes 357

Bibliography 363

About The Author 369

Also By Larry Malerba, DO 371

Acknowledgments

This book would not have been possible without the dedication, support, patience, and kindness of my loving wife Mary.

Special thanks to Don Salmon whose valuable feedback and unselfish assistance encouraged me to pursue this project.

Thanks to Joel Malerba for lending a hand with his artistic expertise.

Many thanks to Alan Trist, Robert Hunter, and the Grateful Dead for permission to use their lyrics.

Foreword

by Don Salmon

I n his previous book, *Green Medicine*, Dr. Malerba demonstrates in compelling detail that our current healthcare system focuses on short-term technical care of the physical body at the expense of our long-term emotional, mental, and spiritual health. His holistic approach goes beyond an examination of the healthcare system itself. It looks at various ways the world we live in negatively impacts our well-being, and it taps into the enormous healing potential inherent in taking into account the infinite spiritual reality of which we are all a part. In fact, he very carefully shows that it is simply impossible to separate out spiritual principles from health and well-being. According to Malerba, a truly "green" therapeutic perspective considers the inner energetic, psychic, and spiritual dimensions of human beings just as important, if not more so, than our external physical bodies.

Although *Green Medicine* is elegant and compelling, Malerba understands that there is a great deal of resistance to the acceptance of holistic, green principles within establishment medicine. He attributes this, in large part, to the fact that the kind of "knowing" that is needed to recognize the inner spiritual dimension is very different from the rational, "left-brain" perspective that prevails in medicine today. A left-brain approach leaves medicine without a framework or worldview within which to address the more complex and subtler aspects of human illness. It leaves no way to account for the innate healing capacity of our bodies, hearts, minds, and souls. And it makes it hard to understand the value of treating the whole person as opposed to simply suppressing symptoms.

In *Metaphysics and Medicine*, Malerba addresses head on the inadequacies of the worldview that currently frames medical science, and offers a very different philosophy that can lead to a new, more comprehensive way of practicing medicine. Don't be thrown by the title. You may see the word *metaphysics* and wonder what something that sounds so abstract and impractical could have to do with medicine. As Malerba clearly shows, it has everything to do with it. So let's look a little more closely at what metaphysics really means.

Simply put, metaphysics is a framework or worldview made up of beliefs, values, and assumptions about what we deem important and what we take to be "real." All the choices you make—from the people you choose to spend time with, to the work that you do, the places you live, to how you use your leisure time—are shaped by what you value as real and important. So what is the worldview, the "metaphysic" if you will, that currently shapes and determines how medicine is practiced? Medicine is deeply affected by the metaphysical perspective of contemporary science, which is, to a large extent, based on the following assumptions:

1. That the universe we inhabit is ultimately a meaningless and purposeless place.
2. That all our cares, hopes, passions, and dreams—the things that determine our choices in life—are nothing more than the result of purely physical/mechanical brain processing.
3. That there is no such thing as free will.

These scientific assumptions exclude the energetic, psychic, and spiritual dimensions that are fundamental components of a green medical approach. You may still wonder whether or how such assumptions actually affect you, either as medical practitioner or patient. As a clinical psychologist who has treated hundreds of pain patients using psychological methods, I offer pain as an example.

What's the first thing a person does when seeking relief from pain? Like most of us, he or she reaches for a pill, perhaps applies heat or cold, gets a massage, seeks out a physical therapist, or looks for some other sort of physical treatment. When I first explain to patients how the mind is involved in pain, they suddenly become metaphysicians, exclaiming, "No, the pain is not *just* 'in my head'... it's REAL!" If you think the same way as my patients, then you, too, think like a philosopher and, yes, you would be doing metaphysics. But your metaphysical reasoning would be in error.

The *International Association for the Study of Pain* defines pain as "an unpleasant sensory and emotional experience that is associated with actual or potential tissue damage." And the definition goes on to say that "pain is always subjective." The *Association* assures us that what is "subjective" is as "real" as anything objective. In other words, our subjective experience plays an important role and has a substantial impact on physical as well as psychological disorders.

But the examples in *Metaphysics and Medicine* go far beyond the issue of pain. An essential component of an effective holistic, non-materialistic approach to medicine is a recognition of the roles played by unusual and potentially powerful inner experiences—such as synchronicities, dreams, and spiritual connections—as well as the inclusion of various non-physical modalities that make use of subtle energies and deep, heart-to-heart, and soul-to-soul communication. While offering a variety of compelling examples and case studies, Malerba leads us step-by-step through an examination of the fundamental metaphysical assumptions that support the mainstream, materialistic, medical mindset. He shows how each facet of these mechanistic assumptions—reductionism, objectivism, rationalism, etc.—leads to the downplaying of our subjective experience and makes it virtually impossible to incorporate the inner, and more fundamental, spiritual aspects of ourselves into our worldview.

What is the root of the metaphysical view that leads us to downplay our subjective experience and the importance of consciousness in general? Malerba's answer, and a running theme throughout the book, is that it is a particular "way of looking" at the mind and body. And this way of looking relates strongly to a particular hemisphere of the brain.

Dr. Iain McGilchrist, a London-based psychiatrist, spent 20 years researching the literature on differences in functioning between the left and right hemispheres of the brain. While these differences remain the subject of controversy, McGilchrist has identified one difference in particular that is supported by a great deal of solid research, and this one difference has a profound effect on the way we view the world.

In brief, the left and right hemispheres have very different ways of attending to the world. The "left-mode" analyzes the world at a distance in order to see it objectively and to gain better control over it. It tends to focus narrowly on the details, "seeing the trees," but at times, "missing the forest." It is less connected to our body and emotions than is the right hemisphere, and tends toward a form of dualistic or "black and white" thinking.

In contrast, the "right-mode" attends to the world with a wider focus, seeing the "forest" or big picture, and experiencing life with a full range of emotions. Rather than detaching from the world, when we employ the right-mode we feel more connected with the world and are more capable of empathizing with other people. While we all employ both modes of attending, most tend to use one mode more than the other.

McGilchrist observes that scientific research over the past several centuries has tended, for the most part, to increasingly favor the left-mode. It is not much of a leap to see that the major "-isms" that Malerba describes—objectivism, reductionism, dualism, rationalism, and so on—are all characteristic of the way the left hemisphere attends to the world. Malerba brilliantly uses the neuroscientific understanding of hemispheric differences to provide us with a practical way of understanding how our materialistic assumptions have so adversely affected the science and practice of medicine.

In the final section of the book, Malerba offers us a wholly different view, one that honors subjective experience and establishes a truly spiritual basis for the practice of medicine. He offers a whole different philosophy of medicine, one that is not only green, but includes the entire "rainbow" of our experience, our world, and the infinite Reality to which we are all inseparably connected. Read this book with an open mind, and you will find yourself questioning every assumption you've ever held regarding the theory and practice of medicine. Prepare yourself for a joyous adventure into the rebirth of medical philosophy, and perhaps a new sense of yourself and the world around you.

****

Don Salmon is a clinical psychologist, composer, and author of *Yoga Psychology and the Transformation of Consciousness*. A grant from the Infinity Foundation enabled Salmon to write a comprehensive study of psychology based on a synthesis of the yoga tradition presented by 20th century Indian philosopher-sage Aurobindo Ghose. He and his wife, Jan Maslow, are currently working on a project that presents mindfulness and related practices in the context of "interpersonal neurobiology." Visit www.remember-to-breathe.org for more information.

Introduction

For quite some time now, freedom of thought has been under siege within the medical profession. More often than not, the war against new ideas is justified in the name of science. Because we no longer give pause to contemplate the ideas that constitute the foundational beliefs of modern medical science, it has been allowed to run off the rails, without the checks and balances that would normally serve to keep it honest. Medical science is on autopilot, hurtling along without the benefit of guiding principles to steer its course. Having lost its way, unable to honestly reflect back upon itself, it is no longer what it claims to be. As a consequence, medical science has become resistant to new concepts, intolerant of freethinking, and is fast falling behind the curve of new paradigm medical theory and practice.

When a discipline like science becomes so certain of itself that it believes it can manage without periodic reexamination of its basic principles, it starts to resemble a doctrine. The more doctrinaire it is, the less receptive to outside input it becomes, and the more it balks at challenges to its authority. Modern medicine has reached a point where its

7

authority is rarely questioned. In the final analysis, medical science is allowed to justify its assertions simply by virtue of the claim that it is science. There is no true accountability to any outside authority other than governmental agencies that increasingly do the bidding of corporate biomedical interests. The one glaring exception is the growing number of individuals who are beginning to question after having experienced adverse medical circumstances in their personal lives.

We romanticize science as man's search for truth when, in actuality, it often plays the role of defender of the status quo. Rather than a tool for further exploration, science has become a means to stifle inquiry and free speech. Science as an ideology becomes an impediment to civilization when that civilization neglects its philosophical heritage, spiritual development, and moral responsibilities. One of those responsibilities is to ensure that no one sector of society is allowed to abuse its authority. The only viable solution to scientific overreach will be a return to freethinking. Just like freedom of religion we need freedom of medical choice—not only the freedom to choose a conventional and/or unconventional form of medical treatment without risk of discrimination, but also the freedom to debate basic principles that underpin the theory and practice of science in general, and medicine in particular. To do otherwise would run contrary to the spirit of science itself.

What was once considered science at the birth of the Scientific Revolution would resoundingly fail to meet the standards of the medical science of today. Most people naturally assume that this is a good thing. Such assumptions are a function of the myth of scientific progress, which posits that science inevitably evolves over the course of time toward greater precision, certainty, and understanding. We take it for granted that the knowledge provided by contemporary science gives us a more accurate representation of reality than the science of bygone eras.

But this is a highly debatable point that philosophers of science have wrangled over for decades. It is true only in the sense that it applies in a specific and limited way to the material dimension of existence. Science focuses its attention on matter but says nothing about the immaterial, that aspect of our lives that involves purpose, meaning, spirit, and soul. It treats emotion, intuition, imagination, and even psychology as ancillary topics. The works of an intellectual giant like Carl Jung, for example, have become a mere footnote in academia because it has become preoccupied with the biological aspects of psychology. As far as neuroscience is concerned, consciousness doesn't even exist, except as a byproduct of brain anatomy and function.

It is also true that modern science has become increasingly imperialistic, overstepping its bounds and staking claim to basic truths that historically have been the exclusive province of religion, theology, and metaphysics. Science has become so overconfident in its abilities that it believes it can dispense with the need for self-reflection. After all, the ground rules of science have been established beyond a shadow of a doubt, or so it thinks. All that remains is for science to put its principles into practice, to map out hitherto uncharted territories until all that can be known will be known. The message is clear: the only reality of import is the hard, cold dimension of material existence. All else is unscientific, insignificant, and has little relevance to human health.

Modern life is becoming increasingly secular. We have become a cynical lot. The idea that a fox crossing my path in the woods could be a representative of the spirit world, sent to convey a message, would be laughable in the eyes of most Westerners. Many American Indians would beg to differ, having been exposed to cultural influences that define Nature as an expression of their creator. Have you ever witnessed or been told a story about a flock of birds flying above, or settling in the trees outside, just at the moment of death of a loved one? Science would have us dismiss our intuitive sense of such events as superstition. Likewise, the sighting of a shooting star is just another astronomical event to science, having no astrological significance or relation to the circumstances of its observer. Western culture embodies values and perspectives that are consistent with its scientific worldview. Modern life is increasingly defined by the tangible, quantifiable reality that science has mapped out for our convenience, drained of all its symbolic, synchronistic, and spiritual meaning. There is little room for magic or imagination anymore.

Our form of medicine is a reflection of the culture that we live in. Human illness, likewise, has become a strictly physical event. There is no purpose to suffering; it is merely an inconvenience, a glitch in the biological program that needs to be overridden. By severing consciousness from disease etiology and development, science believes that it has purged the program of superstition. We can be convinced of this only if we willingly suspend our belief in our own experiences, trusting what medical science has to tell us instead. The more we distrust ourselves, the more medicine is inclined to believe its own version of events. When we open our eyes to compare our personal health-related experiences with what medicine teaches us about illness, treatment, and healing, we begin to realize that something is amiss. It becomes clear that physical medicine is incapable of adequately addressing the deeper causes of illness and the needs of the human psyche, which, in turn, is often at the bottom of our physical complaints.

The modern conflict between science at one pole and religion at the other is a misunderstanding that arises from polarized thinking. Real people are stuck in the middle, their lives hanging in the balance. Each presumes to know what is best for the health of our bodies and souls, respectively, and both tend to dismiss feedback from those that they are pledged to serve. We collectively buy into the legitimacy of these external authorities, no longer believing what personal experience has to offer us. Medical science is quick to point out that subjective experience cannot be trusted. The rational theories of medicine take precedence over the experiential truths of patients.

In my estimation, this conflict between scientific knowledge and personal experience is the crisis of our time. The purported unreliability of subjective experience is one of the primary tenets of scientific method. Science has been waging war against subjective experience for a very long time now—and it has taken its toll on Western culture. In the same way that organized religion makes us hesitant to trust personal spiritual truths, medical science has undermined our belief in the capacity to make health care choices that are in our own best interests. It causes us to doubt first-hand experience and our capacity for discernment. Collectively speaking, we are an immature, ungrounded, and confused people.

It is my contention that, on this particular count, medicine has got it completely wrong. Medical science does not trump individual experience; it should be the other way around. In the end, the only truly reliable gauge of health and wellness is a person's own assessment of him or herself. When push comes to shove and my doctor insists that the drug I am taking cannot possibly be related to the side effect that I am experiencing, I choose to trust my own judgment.

Personal experience should not need validation from science in order to justify itself, although it is certainly nice when the two agree. Medical science, on the other hand, does need to be verified by experience in order to become relevant. If a medical idea, principle, or therapy does not bear itself out in the practical experiences of patients, then one has to wonder about its reliability. A critical examination of the metaphysical beliefs that inform the scientific enterprise will help shed some light on this dysfunctional reversal of priorities. Restoring trust in experiential knowledge will be the first step in rectifying the imbalance created by Western culture's overreliance on left-brain, analytical thinking.

The rise of scientism is a phenomenon unique to our times. This has been made possible by the increasingly materialistic nature of Western culture. Science at its beginning was conceived as a methodology designed to help us learn about the magnificent and mysterious universe around us. Over time, science grew too big for its britches, distancing itself from religion and spirituality to the point where it eventually began to question religious belief. Although the two disciplines are, in my estimation, complementary perspectives, contemporary science increasingly frames its agenda in opposition to religious thought. Modern science, however, fails to comprehend the obvious, which is that is has no ground to stand on when it comes to the larger truths of human existence. It fails to realize that its discoveries have distinct limits—they apply only to the material universe. Because science has deliberately divorced itself from meaning, values, consciousness, psyche, and spirit, it is incapable of achieving a truly holistic understanding of the universe and, of particular importance to this book, of medicine and healing.

Nevertheless, science pushes on, encouraged by a civilization that is dazzled by its technical accomplishments. As its confidence grows, science increasingly fills the void for many who have little familiarity with genuine spirituality. There is good reason why so many have turned to Eastern spiritual disciplines for sustenance—it is because Western culture has become dangerously alienated from its own spiritual roots. Into the vacuum steps science, functioning as a substitute for religion for many, providing a sense of hope and meaning in an otherwise impersonal and materially impermanent universe. When science fulfills this need, however, it ceases to be science. It is in danger of becoming just another competing dogma seeking to supplant other, similarly dogmatic religious perspectives. When science becomes an ideology it is no longer science; it is scientism.

We live in a time when science and scientism appear to the masses to be one and the same. The psychological immaturity and spiritual ignorance of American culture, in particular, has made it increasingly susceptible to the imperialistic impulses of scientism. The tables are on their way to being turned. Whereas religious fanaticism once led to the persecution of heretics who dared to question, it is now conceivable that we are not far off from scientistic persecution of those who challenge the dogmas of science. There is no greater evidence of this than in the medical arena, where ideas proposed by alternative practitioners of holistic medicine are routinely rejected without fair deliberation, regardless of their merit or practical ability to heal the sick and suffering. They are branded pseudoscience the minute they deviate from conventional scientific understanding. In short, they constitute a heretical threat to medical dogma.

Of course, conventional science's lack of awareness of its own metaphysical presuppositions is the very thing that predisposes it to scientistic influence. The only feasible solution is to educate ourselves regarding the differences between science, scientism, conventional medicine, and scientistic medicine. This book is about the philosophical and practical differences between science as it was originally conceived, science as it is construed by mainstream medicine today, the particularly disturbing modern trend called scientism, and a more authentic and inclusive form of future medical science that will no longer ignore the lessons learned and knowledge gained from subjective experience.

What passes for medical philosophy today is really medical ethics, which, although important in its own right, does not concern itself with questioning the unexamined foundational presuppositions upon which the practice of medicine is built. Most books and articles about philosophy of medicine are unadventurous discussions of bioethical dilemmas posed by orthodox medical practice. Basic metaphysical assumptions are never questioned, only the ethics regarding the practice of medicine as it presently exists.

True philosophy of medicine that examines medicine's philosophical roots has no real voice. Academic philosophers of science rarely challenge the Western medical paradigm itself. They content themselves instead with debating the issues that fall within the domain of the medical mainstream. Instead of leading the way as they should, they follow the lead of conventional medical theory and practice. I can only surmise that they are either too thick into it, unable to see the forest for the trees, or concerned about the potential backlash. It is no easy task to question the cherished beliefs of a medical establishment that receives broad support from a culture defined by the same scientific worldview.

Academic philosophers will no doubt find fault with this work on any number of counts. But I am not so much concerned with academic standards of philosophy as much as the practical implications of the ideas expressed herein. I am not and do not claim to be an academic philosopher. I wish to convey these ideas to a wider audience, to those with more than just philosophical or academic interests. I believe it is important for both patient and practitioner alike to become cognizant of the beliefs that lie at the root of Western medical principles. Without awareness of those beliefs, it will be difficult to institute any kind of meaningful change.

The concepts, ideas, and beliefs outlined in this book derive from my twenty-five years of experience as an unconventional medical practitioner with a conventional medical background. It is a pragmatic philosophy developed from the first-hand experiences of patients and empirical observations made by practitioners in the clinic. Evidence was gathered first, philosophical implications were considered later. Philosophical ideas were developed as a response to, and in order to make sense of, the empirical facts.

One of the great mistakes of contemporary medicine is that its philosophy, which is largely unconscious, nevertheless, is the guiding factor that informs its practices. Putting the cart before the horse, medicine looks for evidence to confirm its predetermined philosophical assumptions regarding the nature of illness and cure. It puts theory before practice. When patient outcomes do not corroborate its theories, it persists, forcing the issue in the hope of making reality conform to theory. A simple example of this is the failure of antibiotics to keep up with increasing microbial resistance. Since it is reluctant to revisit the basic principles of germ theory, medicine continues to assume that the only viable approach to infectious disease is to kill bacteria and viruses. This leaves it with no other option but to continue doing more of the same. It virtually guarantees that medicine will remain locked into a framework that is defined by its unexamined presuppositions, and not by its practical experiences with patient outcomes.

The problem with much academic philosophy is its ivory tower lack of relevance to real people and real issues. It is abstract in the extreme, seemingly formulated from thin air, sometimes without reference to any practical matter whatsoever. It is not my intention to disrespect philosophy—in fact I am proposing a renaissance in the field—but, when it comes to medical philosophy, it must be applicable to the actual practice of medicine. This book is intended to be a resource for the development of a practical medical philosophy that addresses the real needs of actual patients and practitioners. It is my attempt to bridge the gaping chasm between the two disciplines in the hope that philosophy will become indispensable once again to the art and science of healing.

Academic philosophy of medicine currently has little influence over the actual practice of medicine. In concurring with medicine's fundamentally materialistic approach to healing, medical philosophy loses an opportunity to make a significant contribution to the advancement of medical science. By deferring to medical authority, it has abdicated its responsibility in thought leadership, thus allowing it to become a subsidiary discipline. Failing to question the foundational principles of medicine, it encourages the continuation of a lazy attitude on the part of the profession, which is disinclined toward any genuine or meaningful self-examination.

Dysfunctional medical practices and poor patient outcomes begin with misunderstandings, with false beliefs regarding the nature of health and illness. Lack of knowledge of its own scientific presuppositions makes it nearly impossible for conventional medicine to develop a comprehensive and coherent theory of disease development and treatment. Without a conscious philosophical framework to guide it, medicine has deteriorated into a haphazard and capricious collection of treatments and procedures that are subject to change at a moment's notice. The latest study contradicts a previous study, which invalidates the original study. This is not, as we are led to believe, the trials and errors of science at work. It is medical science navigating without a map, without awareness or understanding of its own guiding principles.

Reform of the American medical system has been a subject of hot debate for decades now. Many call for universal health care while others fear the erosion of capitalist ideals. Although economic reform of the medical system is a necessary and admirable goal, it will only give us more of the same dysfunctional medical practices that have generated so much iatrogenic illness. Real change can only take place when we begin to rethink the basic principles that inform the practice of medicine. A revolution of understanding is in order.

The modern world appears to be divided into two camps, those who believe in reason and those who give credit to faith, as if the two are mutually exclusive. Western culture is being torn apart by polarized extremes of scientific and religious dogma. The same dynamic applies to medicine. The only way to find a balanced middle ground is

15

through a clear and unambiguous understanding of the issues involved. It should be understood from the start that although this is a strong critique of the current status of medicine, I am in no way opposed to science when it is conducted properly. Good science is aware of its guiding beliefs, understands its limits, recognizes the legitimacy of other disciplines, and respects the multitude of non-scientific sources of knowledge.

In the spirit of full disclosure, I am not a formally religious person. I count myself as one of a growing number of spiritual-but-not-religious individuals. I take my spirituality seriously and I respect the rights of others to pursue their beliefs and non-beliefs. As far as I am concerned, I am a spiritual being temporarily clothed in a material body. I embrace diversity of religious expression. I do not judge others' belief systems, so long as they do not wish to impose their views on me.

I also respect science for the tool of discovery that it is. I have a combined nine years of undergraduate and graduate scientific training. I side with the general consensus regarding climate science and I have no problem with evolutionary theory—as long as it is acknowledged to be theory and not irrefutable fact. I believe that the psychic and spiritual evolution of humankind is contiguous with the physical evolution of the planet and its life forms. I do not view science and spirituality as mutually exclusive. There is plenty of room for both in my life.

However, I am concerned about modern science's pervasive disregard for experiential authority, its increasing disrespect for the sacred, and its general encroachment into territory where it does not belong. The same applies to medicine as a whole—it makes light of patient input by calling it anecdotal, focuses almost exclusively on the physical, and downplays the role of consciousness in health and healing.

Please understand that my criticisms are not to be taken as wholesale indictments of either science or medicine. It is the trend in popular culture to refer to those who critique science or medicine as "science deniers." Many have fallen for this trap, in essence, arguing that science should not be open to criticism. A healthy scientific culture must not only be tolerant of critical review from both insiders and outsiders, it should welcome it. It should also be kept in mind that my criticisms are directed at the theory and practice of conventional medicine and not the individuals who employ it or subscribe to it.

A common debate within philosophy of science is whether science is capable of revealing the true nature of reality or whether it is a tool used to achieve practical ends. My purpose here is not so much to debate the nature of truth or reality. I am more concerned with whether medical science, as it is currently formulated, is suited to its original mission of healing the sick. Unfortunately, science and medicine seem to have become sidetracked by territorial concerns with religion and spirituality, by questions of whose version of reality is true and whose is false. Medicine will have to relinquish its ego, its need to be top dog, so that it can refocus its energies once again toward the practical matter of healing the sick. And it cannot effectively do this unless it first reconsiders its foundational beliefs.

With that said, I will be making reference to *holistic reality* at various points in this book. By this I mean reality as it is, before it has been dissected, quantified, and filtered through the lens of scientific scrutiny. Holistic reality encompasses all phenomena, including consciousness and subjective experience, not just the objective material aspect of human health that medicine deems reality. In practical medical terms, it also means the short and longer-term outcomes of diagnostic and therapeutic interventions, from the perspective of both clinicians and patients. Holistic reality is an inclusive term that leaves out nothing, or at least as little as possible.

Conversely, I will be using the terms *conventional science* and *conventional medicine* to refer to the prevailing collective beliefs and practices of contemporary scientists. As we shall see, those practices and beliefs are defined by a unique set of foundational *–isms* that are not based in science but in metaphysics. Although my focus here is on medical science, I have used the terms *science* and *medical science* interchangeably throughout, largely to avoid repetition. Similarly, I will use *conventional*, *Western*, and *orthodox* to refer to medicine in general. Although the general thesis of this book is meant to apply more specifically to medicine, the arguments made are often, but not always, applicable to both medicine in general and the larger domain of science itself.

I would also like to clarify the meaning of the word, *healing*, as I will be using it. The word commonly carries a fuzzy feel-good kind of New Age connotation, which many assume to be a vague reference to feeling psychologically and spiritually well but having nothing to do with the treatment of "real" physical disease. In medical settings, it often means treating patients with kindness and compassion as they convalesce. This is *not* my definition of healing. Healing, to me, is a process that leads to health on all levels—physical, emotional, mental, and spiritual. Healing is a whole person phenomenon that involves much more than just targeted symptomatic relief, temporary palliation, or forcible suppression. It implies that the overall health of a person is moving in a positive direction over the course of time. This is in contradistinction to a person who, for example, may experience relief from pain after having gallbladder surgery, but who continues afterwards to experience poor health in general. Likewise, a person whose bronchitis improves with antibiotic treatment but who subsequently develops bronchitis the following winter would not qualify as an example of healing. Real healing presupposes that the tendency toward bronchitis will diminish as the overall health of the person improves over time. I will be making the case that our philosophical understanding of concepts like *disease*, *treatment*, and *healing* plays a significant role in the practical outcomes of medical interventions.

I decided to include a glossary at the end of the book because some of the terms that I use have multiple meanings that can be dependent on context and the persons using them. My understanding of empiricism, for example, is very different from orthodox medicine's use of the term. For this reason, I urge the reader to consult the glossary as needed to clarify any misunderstandings that may arise.

Allow me to dispel a common myth that I wish to put to rest right from the start. Most unconventional medical practitioners are not out to undermine orthodox medicine in order to promote their own disciplines. The notion that they do what they do to make money is preposterous; most would be better off financially if they practiced conventional medicine. Many come to their calling after becoming aware of the practical shortcomings of Western medicine. They often endure hardships as a consequence of bucking the mainstream in their quest to provide quality alternative health care to their patients. This book is not about overthrowing the medical establishment. It is about rebuilding medicine from the conceptual ground up, utilizing the best of all medical worlds for the benefit of all.

A recurring theme throughout the book will be the differences in medical perspectives defined by left and right-brain worldviews. Left-brain analytic thought and right-brain holism are broad and imperfect generalizations. Please bear in mind that this is just a convenient way of referencing the issue. It is not meant to give the impression that the different perspectives are mutually exclusive. It is not meant to imply that left-brain thought is strictly rational or that the right brain is exclusively intuitive. Even though most individuals employ a complex mix of the two, many are oriented more in one direction than the other. Since Western societies as a whole are decidedly left-brain in their orientation, they are naturally less concerned about the overreaches of scientific imperialism. For this reason, we must all become fully acquainted with this issue, which threatens to divide people into polarized camps.

Modern medical theory is a holdout from an earlier way of thinking. It is increasingly out of step with the leading currents of our time. Science is historical. It evolves over time because assumptions change as experience changes. The scientific worldview is deeply entrenched in all aspects of culture, including entertainment, industry, the military, government, and medicine. As such, it should require constant reevaluation. Science can no longer dissociate itself from philosophical discourse. After all, philosophy of science originally emerged as a result of the contradictions inherent in the claims of science.

Medicine has two options. It can resist, attempting to prove, disprove, explain, or debunk each new phenomenon encountered on its own terms, according to the principles of conventional medical science.

Or it can relent, allowing the lessons of acupuncture or Ayurveda, for example, to influence its perspective. It can seek to force its conception of the way things should be upon those phenomena, or it can permit its philosophy to adapt to them, amending its presuppositions and principles as its understanding evolves along the way. It can impose its philosophy upon the world or, it can allow new experiences to change its philosophical orientation.

It is time to reestablish philosophy of medicine's legitimate place alongside the practice of medicine. Without it, medicine will be unable to find its way out of the maze of confusion that it has created for itself. Key to that goal is freedom of discourse, the ability to discuss ideas related to medical theory and practice without censorship and without backlash from conventional medical interests.

Part I

The Many *-isms* of Medicine

CHAPTER 1

Medical Presuppositions

Scientists and religious people alike, without exception, put their faith in some belief system that transcends the scope of their present knowledge.
    –B. Alan Wallace, *The Taboo of Subjectivity*

C ontemporary science has achieved a popular reputation for being hard, factual, and almost incontestable as to its conclusions. The term *scientific* is commonly misappropriated by a multitude of parties to imply the truth or unassailable validity of a wide variety of assertions both legitimate and suspect. Advertising claims of products backed by scientific evidence have become legion in the American marketplace. Everything from engine oils to cleaning products, physical fitness fads to dentists' preferences for toothpastes, to the latest pharmaceutical miracles—they all proclaim the backing of scientific evidence. While an astute person understands that marketing strategists will try just about anything to sell products, the clamor to appear to be scientific says something very powerful about the disposition of the Western psyche.

This deep admiration for science is a uniquely modern attitude. It is a form of cultural reverence reserved only for the religious institutions of bygone eras. As I have argued in my previous book, *Green Medicine*, the patriarchal, analytic, left-brain bias of Western culture predisposes it to place undue credence in a scientific mode of perceiving and understanding the world. A left-brain mode of thinking is one that is logical, linear, values certainty, and perceives issues and ideas in either/or terms. In contrast, a right-brain mode of perception is non-linear, intuitive, accepting of ambiguity, and more holistic and inclusive in its approach to problems.

Our deference to scientific thinking is so deeply ingrained that it often borders on a form of dogmatic faith that denies the validity of other possible ways of interpreting the natural world and human experience. It has become commonplace for hardcore proponents of science to silence questioning voices with what they believe to be irrefutable facts produced by scientific research. Most of the time, this type of intimidation works, as dissenting voices usually cannot cite research to back their claims precisely because the scientific community has not previously studied their alternative ideas. More to the point, those voices often fail to understand that they need not frame their arguments according to the conventions of science in order to make valid and intelligent points.

Nowhere is this trend toward unquestioning belief in science of more consequence than in modern medicine. As we shall see, a great deal of medical dysfunction arises from attempts by medical scientists to turn the often messy, unpredictable, and mysterious nature of human illness into a hard, quantifiable, reliable body of scientific information. In essence, the rational medical mind tries to impose its worldview and need for order upon a dimension of human experience that simply is not compatible with its three-dimensional material perspective. Human suffering will always refuse to be pigeonholed in such limited terms. Throughout this book we will explore the reasons why this is the case.

Although modern medicine claims to be based in scientific fact, a careful examination of its underlying principles reveals it to be other than one would expect. The popular conception of medicine as a body of hard scientific knowledge does not square with the everyday realities faced by individuals who rely on the medical system. When we look more closely, we find that patient experience is often at odds with medical fact. The doctor says one thing, the patient says another, and it becomes clear that the two perceive and experience reality from different frames of reference.

During my years of medical training my professors never once broached the issue of basic assumptions that underlie the conventional medical worldview. As basic assumptions I can only surmise that they were taken for granted to be true. Students were fed the facts and expected to regurgitate them as gospel without acknowledgment of the philosophical principles that underpin those facts. We were neither taught nor encouraged to *think* about medicine.

The foundational principles that inform the medical enterprise tend to go unexamined by the vast majority of medical personnel and practitioners. For many, those principles are unconscious. Although it usually requires an act of will to maintain such a state of unfamiliarity, it is understandable why these important issues would go unnoticed for so long. To bring them to awareness would potentially generate much cognitive dissonance and a good bit of personal discomfort.

How odd it is indeed that the assumptions we make about the world we live in, what it is to be human, and the nature of health, illness, and healing would not generate more interest among the medical profession. Medicine would prefer, it seems, to skip that one very important first step in order to get right to the "facts." That first step involves the definition of terms and formulation of principles that constitute the foundation upon which to base our medical knowledge and our strategies regarding health and disease.

An overreliance on conventional medical beliefs predisposes one to an unnatural skepticism toward that which cannot be seen, felt, heard, measured, or quantified. As a result, a great deal of human experience is eliminated from the medical equation. By contrast, considerations regarding the essential nature of health and illness have become the primary focus of many holistic green practitioners who understand that one's foundational assumptions play an important role in treatment options and their consequences.

Medical science is averse to looking too closely at its own assumptions precisely because they involve metaphysical considerations. Metaphysics is a broad term that is used here to refer to a branch of philosophy that examines first principles. Those first principles often take us to places that transcend the strict materialism that undergirds conventional medical beliefs. Although metaphysics is thought to be incompatible with true science, there would be no science at all without the metaphysical assumptions that form the foundation of its theories, methods, and conclusions.

Aristotle called metaphysics the "the queen of the sciences" because, to him, it was first in importance. It is a branch of philosophy that encompasses all of reality, both objective and subjective. In fact, it is not so much a branch as it is the root of all philosophy. In times past, it was considered a science, a science of universal principles pertaining to all of existence. Modern science concerns itself with the specific physical details of material reality. Having been displaced by modern science, which is now believed by many to be the arbiter of truth, metaphysics is no longer considered a science, and is taken to mean the philosophy of the world beyond nature. In short, it is now considered the philosophy of the immaterial.

Materialism, by definition, obviates the need for a metaphysical perspective because it does not acknowledge the existence of an immaterial reality. Modern medicine, being fundamentally materialistic in nature, chooses to ignore deeper questions of metaphysical concern, thus guaranteeing that it never changes its orientation toward issues critical to the art and science of healing. Questions like "what is the nature of life" "what is matter" "what is spirit" "what constitutes causation" "what is health" "what is illness" and "what is healing?" are metaphysical questions.

While medical science trains its attention toward the facts of physical existence, metaphysics contemplates the broader principles that apply to life in general. It is not possible to have one without the other. They are complementary disciplines, with science trending toward reductionism and metaphysics moving in the direction of holism. Metaphysics has the potential to unify the vast and fragmented world of scientific specialties and sub-specialties. Although science believes that it can exist without metaphysics, that, in my opinion, is just a form of self-delusion. A materialist doctrine that lays material claim to all of reality is a metaphysical perspective whether it is acknowledged or not. It is a statement about the nature of the universe that we experience and inhabit. While it is just one among many possible metaphysical positions, it does happen to be the predominant perspective embraced by contemporary scientific culture.

The irony is that no system of healing has the luxury of resting on a hard factual foundation because most foundational assumptions are, by their very nature, metaphysical. Science wishes to project an image of certainty, which it attributes to the purity of its methods of fact finding. This line of logic fails to hold up, though, when we realize that science's most fundamental premises are scientifically unverifiable. Those premises involve non-scientific assumptions regarding the nature of the material world, the non-material world, and human existence's relationship to those worlds.

Unbeknownst to most medical scientists, the entire enterprise of conventional medicine begins with *a priori* metaphysical assumptions that are based neither in science nor empirical fact. Scientific reasoning, therefore, is critically dependent upon what most scientists would characterize as non-scientific metaphysical presuppositions. By a clever twist of logic it would be eminently reasonable for me to claim that the very foundations of science are, in fact, unscientific. But I will refrain from being so severe in my judgment because I realize that we are dealing here with different frames of reference and relative degrees of certainty.

The foundations of science are far from a settled matter. This is especially true of medical science, which deals with a topic that is more complex and multidimensional than other branches of scientific inquiry. Thanks in great part to Thomas Kuhn's contributions to the field of philosophy of science, we understand that older scientific discoveries were not always errors in thinking—as the typical student of science is inclined to believe—as much as they were conclusions drawn from different assumptions made in different cultural contexts during different historical periods.

Science and the knowledge that it produces evolve over time, largely because fundamental assumptions and theories based upon those assumptions change. According to Kuhn, scientists practice "normal science" by interpreting the data within an agreed upon framework of assumptions that they assume to be true. However, scientific revolutions are fueled by those who think outside of the box. New scientific paradigms represent quantum leaps in the way we define reality and the principles that govern it. Simply put, the same old data is seen in a new light. Normal science as practiced by the medical establishment naturally tends to resist change. Nevertheless, there is a green medical revolution that has been quietly underway for quite some time now, and its presence is indicative of the need for a long overdue change in perspective.

The newly emerging medical paradigm defines reality in significantly different terms than the orthodox medical paradigm. Let us take, for example, one of the more important unspoken principles of conventional medicine, which assumes that because we live in a material world, our ailments should, likewise, require material interventions and material solutions. Many in the medical field would object to the characterization that they are materialists, pointing to the fact that they have religious and/or spiritual beliefs, and I would not doubt the sincerity of their claims. The problem, however, is that the form of medicine that they practice does not honor or support those beliefs, and is not compatible with a worldview that acknowledges the reality of a non-material dimension.

If a particularly brave individual were to decide to incorporate spiritual considerations into his or her medical practice, it would have to be undertaken against the grain of the medical monolith. It would be done at the risk of one's reputation, exposing oneself to the possibility of being labeled with any of a number of commonly used medical swear words such as "unscientific," "pseudoscientific," or "quackery." As a practical matter, most practitioners learn to quietly compartmentalize their lives, thinking like scientists in the halls of medicine while keeping their spiritual beliefs confined to home and their houses of worship.

More important, lest we forget, patients themselves have learned through their interactions with the medical system to stick primarily to the "facts." They tend to refrain from making personal observations and expressing their personal beliefs once inside the consulting room. Sadly, this programming to conform to scientific expectation has taught many to distrust their own experiences and judgment. It also deprives the medical profession of pertinent information that could teach it a thing or two about the relationships between human nature and human health.

Conventional medical science is a strictly material enterprise. Medicine assumes that material reality is the only reality of importance. Even when it does on the rare occasion acknowledge the existence of an immaterial dimension, it is thought to be of no practical medical use because most believe that it cannot be controlled or manipulated by the methods and tools of science. The point is that metaphysical assumptions made about the nature of the experienced world and our lives in it have profound implications for the type of medical views that we endorse and support. If I am a strict materialist then I should have no qualms with medicine as it is practiced in its present-day form. If I believe otherwise, then there are many questions to be asked and many issues that remain unexamined by the medical mainstream.

Whether I assert that physical reality is the only reality or that I am certain of the existence of a spiritual dimension—neither of these contentions can be proven or disproven by the standards of proof demanded by rational science. Such assertions are either theoretical or based on subjective experience and cannot be characterized as fact as science would define fact. The assumptions of either camp are objectively unverifiable, which is why they constitute *a priori* principles that must be taken at face value. No concrete evidence can be provided to definitively support one conclusion or the other. This is not to say that there are not other forms of evidence, but that is the type of evidence that science, in keeping with its foundational principles, does not acknowledge.

So we have a bit of a conundrum—or paradox if you will. Science has to take its starting point from somewhere if it is to ever get off the ground. But while most presume that science is based in hard fact, scientific reasoning must always proceed from assumptions—from principles, axioms, and hypotheses that are assumed without objective evidence to be true. One would think, therefore, that it ultimately comes down to a matter of opinion. Either I choose to "believe" that material reality is the only reality of medical relevance or I do not. But as we shall see, there are, I would contend, other ways of knowing that science simply refuses to accept as valid. Although these other forms of knowing may not have the backing of conventional scientific evidence, they are the product of millennia of accumulated human experience.

The inability to envision anything medically relevant but the material is a function of just one of a number of self-imposed biases inherent in the foundational principles of Western medical science. In other words, medical science presupposes, among other things, nothing but the physical and, as a consequence, defines the world that it moves in as strictly material. It is worth noting the circularity of this assumption. A related axiom presumed to be true, particularly in medical science, is the belief that there is one best way to proceed that will always prevail. This assumption lies at the heart of what many patients experience as the frustrations and failings of cookie-cutter medicine.

To insist on one particular way as the only correct way involves making some very big and erroneous assumptions. This, in turn, encourages medical hubris and intolerance toward dissenting opinion. It tends to lead down a perilous path where principles and guidelines become hardened into "–*isms*." Materialism happens to be just one of multiple –*isms* that I will be discussing, especially as they pertain to the practice of medicine. While –*ism* is a suffix befitting of an ideology, doctrine, or prejudice, it is not compatible with an open-minded approach to scientific inquiry. An important motivating factor for writing this book is to highlight the basic principle that while there is one common goal—health and healing—there are many ways to get there.

However, neither I do subscribe to absolute relativism. I do not mean to imply that all methods are to be viewed as perfectly equal. Some are safer and some are riskier. Some are more effective and some are less. Some are about outcomes while others are about process. Some produce local results while others have a more global impact. Some are suited to certain situations and others are not. They all require personal judgment and oftentimes professional assistance in order to be properly implemented. They all have value and should be understood as the choices that people must be allowed to make—for better or worse—on their healing journeys.

It is also true that the assumptions that underpin a given medical methodology are of critical importance to the results that it is capable of achieving. Different foundational assumptions will yield different outcomes. When medicine focuses only on the material dimension of human illness it yields mostly physical results and, I would add, short-term superficial results. If modern medicine were to widen its perspective there is no telling how much more satisfying those outcomes and how much more fulfilling the process could be.

It is imperative for both patients and practitioners to become cognizant of the mostly unspoken beliefs that define the conventional medical care that they have chosen. After all, a shaky foundation is likely to provide insufficient and unreliable support for our house of healing. Let us now examine some of the more prominent and consequential principles that Western medicine assumes to be true and necessary in its approach to healthcare. When elevated to the status of –isms, these assumptions become problematic impediments to medical innovation and successful care of the sick.

CHAPTER 2

Medical Materialism

Nature is not matter only, she is also spirit.
    –C. G. Jung, *The Collected Works*

The day Science begins to study nonphysical phenomena, it will make more progress in one decade than in all previous centuries of its existence.
    –Nikola Tesla

The Scientific Revolution of the sixteenth and seventeenth centuries was at least partially motivated by science's desire to distance itself from religion and "superstition" in search of natural as opposed to supernatural causes. The fallout from this unnatural divergence has had an enormously dysfunctional impact on modern medicine, which has thoroughly assimilated this message as indicated by its continuing belief that the vast majority of our physical, mental, and emotional woes can and should be assigned material causistic explanations.

While the term *science* was originally used as a synonym for *knowledge*, the word *scientist* did not emerge until much later in the nineteenth century. In antiquity, philosophers, mathematicians, and physicians were the analogs of modern-day scientists. They were usually well-rounded, scholarly persons knowledgeable in the arts, sciences, philosophy, and theology. *Natural philosophy* was understood to be the study of nature and the physical universe. Unlike contemporary scientists, natural philosophers acknowledged intangible factors and considerations in their study of nature. They were more attuned to a holistic conception of nature and did not shy away from bigger questions regarding the relationships between material existence and higher cosmic forces. The first person ever appointed to the position of Professor of Natural Philosophy was Italian scholar, Jacopo Zabarella, in 1577 by the University of Padua.

Sir Isaac Newton (1642-1727), considered by some to be the most influential scientist to have ever lived, was an astronomer, alchemist, physicist, mathematician, natural philosopher, and theologian. Although his *Principia* is thought to be one of the most important scientific books ever written, it is estimated that three quarters of his writings were metaphysical documents about the intangible nature of God and the universe. In Newton's time it was not considered bad form to seek a broader understanding of creation that was inclusive of both the heavens and the natural world.

Just as natural philosophy was a later branch of the larger field of philosophy, *natural science* emerged as a subdivision of natural philosophy. Natural science represented a more stringent interpretation of the natural world to the exclusion of intangibles and metaphysical matters. It sought to explain the material world in material terms, further distancing itself from religion, theology, and even philosophy. In time, the subject matter of natural science came to be identified as the accepted body of scientific fact, while metaphysics was believed by "men of science" to be a superstitious form of magical thinking. Most modern universities have schools of natural science under which departments of biology, chemistry, mathematics, physics, and the like are subsumed. Topics like metaphysics, spirituality, and consciousness studies are most unwelcomed subjects in academic circles of science, where they usually view themselves as the guardians of "factual" matters regarding the "real world."

It wasn't until 1834 that the term *scientist* was first proposed by

William Whewell, a Cambridge University historian and philosopher of science. I find it fascinating that Whewell may have suggested the term in response to his perception of the growing division of the sciences into distinct disciplines. He wrote of "an increasing proclivity of separation and dismemberment"[1] among the sciences, whose practitioners had titles such as chemist, astronomer, and physician. He pointed out that natural philosopher no longer served as an inclusive term for the various practitioners of science and suggested that perhaps scientist would be more appropriate. Thus, scientist could be used to indicate one who is concerned with knowledge of the natural world, which was to be understood as distinct and separate from philosophy and the supernatural world of metaphysical knowledge. Right from the beginning, the term was borne out of a perceived need to unify a field that was already trending toward fragmentation.

The historical progression can be traced, therefore, from the more general to the more specific—from learned men of broad education in art, science, and religion, to philosophers, to natural philosophers, to natural scientists, and ultimately to our modern day scientists. The clear trend is toward increasing specialization and away from anything that can be construed by science as speculative. Religion, theology, spirituality, metaphysics, and philosophy gradually came to be seen as impractical and, in the end, irrelevant endeavors that were unbecoming of serious scientists. This bias in favor of hard, reproducible, "factual" matters has reached its pinnacle in contemporary Western cultures where it has taken on the form of an institutional rigidity and intolerance that is particularly characteristic of modern medicine. It strikes me that because human medicine is the least hard of the hard sciences it motivates the profession to strive mightily to achieve the status of a hard science. The irony, again, is that while medical science projects an image of fact-based certainty, its very foundation is predicated upon unspoken metaphysical beliefs.

The net result is a closed system called natural science, which is only interested in the natural world as narrowly defined by material existence. Furthermore, all explanations and causes must, by definition, be material causes. Other non-material explanations and causes are not possible because they have been predetermined to be either non-existent or beyond the purview of natural science. Many scientists believe that all answers are to be found in the natural world and that physical phenomena cannot be influenced by or explained according to otherworldly principles. Medical science can be classified as a subdivision of natural science and, as such, tends to be constrained by the same self-imposed blinders. It is concerned only with the human *body* and the material causes of physical diseases that afflict that body. It is clear that modern medicine has gotten itself into a bit of a box. It is worth asking what the multi-talented Isaac Newton would have thought about this state of affairs.

To make matters worse, most Westerners are so well conditioned by their materialistic cultural milieu that they tend not to question the premises of the dominant scientific forces that shape their lives. And when scientists' plans go awry and events spiral out of control, the majority are puzzled as to why, largely due to the fact that they have not yet learned how to ask the right questions.

It is sometimes instructive to examine language and its various subtleties. We recognize *material* as a reference to physical reality, matter, that which is of substance. Conversely, we know *immaterial* to mean non-physical, intangible, ethereal, or supernatural. But the two terms have other meanings that are, no doubt, strongly colored by the Western mindset. In a court of law, the same terms are frequently employed to convey relevance and irrelevance. Legal evidence that is material is important, pertinent, and essential, while immaterial evidence is inconsequential, unimportant, and irrelevant. The same terms could be transposed to the affairs of medical science and, in most cases, nary an eyebrow would be raised.

If it is not already clear, I should explain. By materialism I mean more than just a mere love of material things and desire for material possessions. It signifies an overemphasis upon the importance of material existence and the physical body combined with a neglect or even denial of consciousness, spiritual principles, and matters of the soul. I believe this linguistic example to be indicative of the misplaced values of our cultural conditioning.

The Western mind has come to define nature as separate from the heavens—independent from and unrelated to spirit, soul, the idea of a higher power, and even mind. The natural consequence of such a dualistic frame of reference is a tendency to live our lives in compartmentalized fashion. Work is separate from home is separate from worship is separate from recreation, and so on. We begin to feel inhibited about mixing ethical issues with business, political and religious beliefs with social life, and spiritual concerns with our medicine. Science in general and medical science in particular have sidestepped the moral and spiritual implications of their actions by claiming that they are inherently neutral, neither good nor bad. This might be true if science were conducted in a vacuum, on a planet where there were no people or any other life forms for that matter. But science is conducted by people and its primary impact is upon people and our planet, which sustains all life forms. Thus, "value neutral" becomes a euphemism for avoidance of responsibility for the ethical and moral implications of the actions we take and cultural conventions that we condone.

In essence, science has separated nature from the gods and greater cosmos and placed it in its sterile laboratory where it operates according to an artificially constructed set of rules that define the limits of the scientific universe. When ideas and influences threaten to encroach from outside the lab, they are dismissed as imaginary, naïve, irrelevant, and unscientific. Mother Nature is thus stripped of her cosmic significance and the Earth is rendered into a ball of rock floating in space. Nature is dissected and studied, and her gifts are engineered, exploited, and sacrificed in the interests of capital and progress. It is no coincidence that this secularization of nature has been collectively perpetrated by patriarchal cultures that place higher priority on left brain, masculine values. It is also no coincidence that this same fundamentally secular orientation dominates medical education and medical practice and is driving the modern biotechnological revolution.

One need only reflect upon the use of the terms *natural resources* and *natural health* in order to grasp the contrasting meanings. The only way to systematically exploit nature without experiencing unwanted pangs of conscience is to convince oneself that this planet and the lives that depend on it consist, at bottom, of nothing but inert matter adrift in a random meaningless universe. Thomas Merton, arguably America's greatest Christian mystic, expresses the sentiment quite well:

> *Technology can elevate and improve man's life only on one condition: that it remains subservient to his real interests; that it respects his true being; that it remembers that the origin and goal of all being is in God. But when technology merely takes over all being for its own purposes, merely exploits and uses up all things in the pursuit of its own ends, and makes everything, including man himself, subservient to its processes, then it degrades man, despoils the world, ravages life, and leads to ruin.*[2]

Matter, according to science, is the "substance" of the natural world. Science has, in effect, carved out the domain of matter as the exclusive subject of its investigations. The immaterial is either of no consequence, does not exist at all, or is simply considered outside the box of natural science. This is an illustration of a concept from philosophy known as the *closure principle*. It is the belief that the material universe is causally closed. According to the closure principle, there can be no causes of physical phenomena other than physical causes. This principle can be traced to the influence of Descartes and has had a profound impact upon the trajectory of science ever since. Descartes legacy of mechanistic matter in motion separate from any supernatural influences is a notion that has infected all branches of the life sciences. Life, as a consequence, has been reduced to the academic study of matter governed by the laws of biology, chemistry, and physics.

Of course, the closure principle also represents one of those arbitrarily drawn conclusions made for the convenience of scientific purposes. In other words it, too, is a metaphysical assumption that has no grounding in scientific evidence. It is an essential component of any materialistic worldview. By extension, if physical phenomena are influenced only by other physical phenomena, then meaning and purpose can play no role in matters of the material world. But if supernatural influences can neither be proven nor disproven by scientific standards, then it would seem to be a matter of personal preference whether one claims that there can be immaterial causes or that there are only physical causes.

This is not to say that it is not useful for science to focus its activities on studying the physical world. Much valuable information and many practical applications have resulted from scientific investigation of the world of matter. Scientific achievements of this type, however, are by definition limited to matter and its physical causes. When medical science mistakenly applies this myopic perspective to human life, health, and illness, the result is a sad comedy of errors that has real consequences for real people. Theories may be constructed and experiments conducted within the parameters of the scientific box but, from the perspective of holism, the end result must have a connection to the larger scheme of things if it is to have a true and lasting positive impact on the quality of our lives.

It is hard to deny the immediacy of the world of physical matter. It is the solid stuff that we encounter on a daily basis. It is the pen that we write with, the chair we sit in, the car that takes us places, and the food that we eat. The irony, though, is that matter itself has come under fire from the very scientific world that has always assumed it to be the one true reliable reality.

While it can be useful to speak of atoms and molecules as building blocks of matter, our conceptual understanding of these entities has evolved rapidly in modern times. Once believed to be the smallest component of matter, an atom was subsequently determined to be composed of a neutron, electrons, and protons. Chemistry has charted the various types of atomic building blocks in the Periodic Table, making careful note of their positive and negative charges and atomic weights. Electrons were said to be orbiting around a central nucleus, not unlike the planets revolving around the sun. The measurement of these microscopic "particles" certainly lent powerful credence to the notion of them as tiny bits of matter—the very stuff of which the material world is composed.

But it turned out that this conceptualization of atomic structure was an oversimplification. Along came physics and Werner Heisenberg who formulated the uncertainty principle of quantum mechanics. In essence, Heisenberg declared that we can never really pinpoint any one of these electron planets in their orbits—we can only speak of their locations in terms of mathematical probabilities. In fact, the more physicists tried to isolate any given particle in order to study its properties, the more elusive it became.

Just a few years earlier, Albert Einstein had proposed his theory of special relativity. From this he formulated his famous mass-energy equivalence equation, $E = mc^2$. Particulate atomic matter was no longer seen as the immutable fundamental unit of physical existence because it was now believed to be interchangeable with energy. It was determined that particles can be created and destroyed and, more importantly, that matter is transmutable into potential energy and various types of radiation.

Advancements in physics continue to alter our once certain conviction regarding the solidity of organic and inorganic matter. Matter itself, the very substance of science, has become a slippery concept indeed. Our definition of matter has, in fact, become increasingly elusive and has undergone numerous revisions, most of which are too abstract, mathematically dependent, and complex for the majority of us to adequately grasp. Suffice it to say, it has become quite reasonable, even necessary, for us to question the physical nature of material existence.

Are we mere physical bodies that will someday return to dust after we die? Is the human brain just a highly complex physical computer composed of millions of neuronal connections? Are our emotions, dreams, desires, intuitions, and psychic experiences mere illusions generated by that slab of beef in our heads? Is material existence the only reality? Is it possible that our scientifically predisposed civilization is mistaken regarding the true nature of life on planet Earth and of life after we leave it? Is it fair to make the broad generalization that the physical universe is the only dimension deserving of scientific investigation? And, what impact do the answers to these questions have upon our approach to medicine, illness, health, and healing?

As a practical issue, matter as a concept is crucial to the activities of science, but it has some serious limitations. Much of inorganic science has been highly successful while staying within the confines of its materialistic assumptions. But medical science is poorly served by such a narrow framework, mainly because it involves that phenomenon we call life. Like the proverbial ostrich, medical science can continue to act in accordance with the closure principle, which is to say that human illness can be studied strictly in physical terms—or it can pull its head out of the sand and take a good look around at the remarkable paradigm-shattering work being done by countless dedicated unconventional healthcare practitioners and their patients.

Although his theory of "immaterialism" represents perhaps the polar opposite of materialism, Irish philosopher, George Berkeley, nevertheless made an important point when he said that matter is a metaphysical concept—a far cry from the solid stuff of the material world that most moderners believe it to be. After all, no one has ever seen matter. What we perceive with our senses is more accurately characterized as the attributes of matter—qualities such as shape, color, mass, and texture. Berkeley argued that we do not perceive matter directly; rather, it is a mental construct, an abstraction based in metaphysical assumptions.

Another philosopher, French metaphysician, Rene Guenon, has referred to science divorced from spirit as "profane" science. He argues quite forcefully—and most spiritually-inclined persons would have to agree—that matter as the only reality is a naïve illusion:

> The truth is that there is really no 'profane realm' that could in any way be opposed to a 'sacred realm'; there is only a 'profane point of view', which is really none other than the point of view of ignorance. This is why 'profane science', the science of the moderns, can... be justly styled 'ignorant knowledge', knowledge of an inferior order confining itself entirely to the lowest level of reality,

knowledge ignorant of all that lies beyond it… Irremediably en-
closed in the relative and narrow realm in which it has striven to
proclaim itself independent, thereby voluntarily breaking all con-
nection with transcendent truth and supreme wisdom, it is only
a vain and illusory knowledge, which indeed comes from nothing
and leads to nothing.[3]

One of the more problematic aspects of Western medicine is its inability to come to terms with the non-physical. Even though solid matter as an idea has been exploded by science itself, medicine continues to
be a matter-based practice. Conventional medicine seeks to define most
health issues as physical problems with material causes that require material solutions. It routinely tries to force the multidimensional nature
of human health experiences into a strictly materialistic framework. An
example of this is the field of psychiatry, which has become increasingly and inordinately influenced by the biological school of thought.
As a consequence, thought, emotion, and issues of spirit have been
marginalized as secondary considerations because they are believed by
many scientists to be mere byproducts of brain electro-chemistry.

Those who suffer from chronic pain are often subjected to numerous diagnostic tests in a search for physical causes to their woes. Material causes such as torn cartilage, nerve impingement, or tissue inflammation are the preferred explanations. Although some cases of chronic
pain can be attributed to physical causes, many have more complex etiologies that lie in the mysterious recesses of the human psyche. In addition, as we shall see, all illness, regardless of its etiology, has an energetic
component that cannot be accessed through biochemical, physiological,
or structural means.

I once saw a woman who had developed a case of shingles (herpes
zoster). The right side of her face was noticeably red, inflamed, and
swollen. Conventional medicine attributes herpes zoster to nerve inflammation that results from reactivation of the chickenpox virus (varicella), which had previously been dormant in the nerve root associated
with the physical distribution of the particular case of shingles—a very

convincing and satisfying material explanation indeed. However, it is also acknowledged that shingles is often a stress-induced disease. Nevertheless, orthodox medical emphasis is typically focused on physical causation and physical solutions. Accordingly, my patient's regular doctor had prescribed both an antiviral medication for the supposed viral cause, and prednisone, a powerful corticosteroid, for the presumed inflammatory cause.

This woman had consulted me in the hope of an alternative solution mainly out of concern for the side effects of these medications, which she refused to take. My investigation included a discussion of potential stressors that may have triggered her shingles. She noted that she had recently been grieving over the loss of two pet dogs but she was not sure that this was enough to bring on her illness. I prescribed a homeopathic medication in lieu of the steroid and antiviral and urged her to call me with a progress report. She called a few days later with a very interesting story. The day after I had seen her she became quite agitated and got into a verbal fight with her husband. Her anger and shouting then gave way to a torrent of tears as she came to the realization of what had been bothering her.

It was suddenly clear to her that she had become very angry and even depressed ever since the political party that she supported had suffered a resounding loss in the recent election. This had compounded the grief she felt over the loss of her pets. Normally, it can take weeks and even months for a case of shingles to fully resolve. Amazingly, the redness, swelling, and inflammation of her face began to recede as she released the cathartic anger and tears of grief. After a couple more days, the physical signs of shingles were nearly undetectable and within a week she experienced complete recovery.

To define this case of shingles in strictly material terms would be very tricky business indeed. Are we to believe that varicella is the actual cause or merely an associated phenomenon? Is inflammation a cause or symptom of the condition? And are these material phenomena the real causes of the illness or are the non-physical grief and anger deeper sources of the disturbance? There is often much more than meets the eye when we carefully examine individual medical circumstances.

Perhaps one of contemporary medicine's greatest blind spots is its certainty as to the material causes of disease. It prides itself on being able to explain the biochemical and/or physiological mechanisms that lead to many illnesses. Such explanations, however, are presumptions grounded in materialistic bias. Regardless of medicine's self-imposed refusal to acknowledge a non-physical dimension, all presumed physical causes can, from a more holistic green perspective, be understood as surface manifestations of deeper energetic, psychic, and spiritual phenomena.

Those who subscribe to a materialistic medical worldview *believe* that the physical is the only reality of relevance, or that even exists. Medicine is so uncomfortable with the non-physical dimension that it artificially excludes it from medical consideration, often with the rationalization that it is unscientific and unworthy of investigation. Emotional, moral, spiritual, and energetic factors as they relate to health are often dismissed simply because they cannot be verified by materialist standards of proof.

CHAPTER 3

Medical Reductionism

Boundary lines, of any type, are never found in the real world itself,
but only in the imagination of the mapmakers.
–Ken Wilber, *No Boundary*

(O)nce the Scientific Revolution had established the independence of physical existence from supernatural influences, science was then free to explore the material frontier without regard for the broader implications of its actions. It conducted its explorations primarily by processes of division, dissection, and separation. The first degree of separation was from the effect science was to have upon human lives. Shielded by the fantasy that science was an amoral enterprise, it could venture wherever it wished without taking ownership of the complex problems that it left in its path. Thus it was an easy step from materialism to reductionism as men of science studied the natural world, broke it down into its constituent components, and attempted to locate the secrets that would explain the nature and function of all things organic and inorganic.

In keeping with its unexamined foundational principles, medical science has historically taken a reductionist approach to investigating issues of human health. It dissects the human body into its component parts and studies them in isolation from all other parts. Although much valuable medical information has been learned from this limited application of scientific method, when taken as the primary approach, the end result is a scattered array of body parts and specialized fields of medicine, all of which are increasingly dissociated from one another.

Medical science is perpetually searching for the least common denominator in its quest to conquer disease. It is deeply invested in locating that one causative material factor that will explain an illness or the one treatment that will cure most, if not all instances of a particular disease. A thorough understanding of the parts, it is believed, will inevitably lead to mastery of the whole. Reductionism creates the illusion of control by limiting the range and scope of scientific activity. It operates within a self-imposed bubble that relieves it of responsibility for events that occur outside the bubble. This is the essence of reductionist thinking. It seeks to impose simple solutions upon seemingly local problems that should require more sophisticated, broad-minded approaches.

Furthermore, the reductionist influence has also led to the belief that treatment of the part not only has no real impact upon the greater whole, but is more important than the whole. In this way, for example, side effects that can impact the person as a whole are minimized for the sake of the desired effect that a drug has upon target symptoms.

Author, philosopher, and noted historian of religion, Mircea Eliade, believes that reductionism is a particularly Western affliction that separates us from higher values that define humanity:

> The explanation of the world by a series of reductions has an aim in view: to rid the world of extra-mundane values. It is a systematic banalization of the world undertaken for the purpose of conquering and mastering it. But the conquest of the world is not—in any case was not until half a century ago—the purpose of any human societies. It is an idiosyncrasy of Western man.[4]

Reductionist thinking also makes it easier to maintain the illusion of control when only a fragment of the larger picture is taken into account.

Removal of a diseased gallbladder (cholecystectomy), for example, is deemed a success when it eliminates a patient's attacks of abdominal pain. It is rightly a success in that limited sense, but the overall impact of the intervention on the patient's subsequent trajectory of health and illness over the course of time is never examined. The cholecystectomy is thus dissociated from its longer-term consequences for the whole person. Medicine prefers it this way because it lacks the holistic understanding to grasp the implications of such events.

Much of contemporary medical science moves in a direction opposite to that of holistic reality. The scientific community is biased in favor of seeking answers by examining the smallest of parts, which they believe will someday yield the keys to health and illness. Awash in a vast sea of scientific detail, Western medicine has lost the capacity to see the big picture, repeatedly failing to connect the dots between medical events. The reductionist methodology of medical science does not readily lend itself to studying the human organism as a whole and, not surprisingly, leads to the conclusion that a holistic approach is not a worthwhile avenue of endeavor.

Two of the greatest achievements of Western medicine, germ theory and genetics, are emblematic of the reductionist quest for final answers. Germs and genes are the tiniest of parts, so tiny that they cannot be seen by the naked eye. Future aspirations for humankind's battle against disease, so we are told, are wrapped up in unlocking the secrets of these microscopic entities. Held out to the general public as the great unfulfilled promises of scientific civilization, the medical establishment would forever have us believe that microbiologists and geneticists are moments away from discovering that crucial bit of information that will lead to a therapeutic breakthrough. And yet each answer just seems to yield unanticipated problems and more riddles. Undaunted by setbacks, medicine encourages us to be patient because it is just a matter of time before science figures it out. While we wait for those unrealized promises, biological weapons, superbugs, genetically modified frankenfoods, and the patenting of genetic entities for corporate gain are just a few of the horrific consequences that have resulted from the pursuit of reductionist science divorced from holistic reality.

A solid argument can be made that the overemphasis on germs and genes has been more detrimental than helpful. Clearly there are benefits, but the scientific community demonstrates a general lack of judgment and perspective when it plunges ahead, as it often does, in disregard for the danger signs. The rise of biotechnology is predominantly a function of an unholy marriage between medicine and corporate interests. The biotech industry relentlessly tinkers with the genetic code as it seeks to develop new genetically modified commodities, the dangers of which are repeatedly minimized. After all, we are told, science is value neutral and must be allowed to proceed unfettered. But the dangers are very real. No human being has the wisdom or knowledge to foresee all of the potential repercussions prior to the creation of a genetically modified organism. And it is impossible to put the genie (pun intended) back into the bottle once the mad experiment has gone awry.

Given that the capacity to alter the genetic code is in its relative infancy, scientists have yet to anticipate the fierce backlash that will likely be coming from Mother Nature. The rotten fruits of our failure to heed the warnings are already apparent, however, in the war against our microscopic neighbors, germs. Most attempts to eradicate these tiny microbes have backfired as evidenced by the development of so-called superbugs that are increasingly resistant to our vast arsenal of antibiotics. Our pharmaceutical weaponry has increased the pressure, thus accelerating the process of natural selection among bacteria with which we might otherwise have had a mutually beneficial relationship.

The war against viruses is also relatively new and common sense indicates that these organisms, too, will outwit our best antiviral ammunition. Because viruses are so quick to adapt and so easily spread, we may be in for a rude awakening when it comes to the long-term consequences of such an unwarranted strategy.

Widespread devastation is already being wrought by biotechnology as evidenced by our scientifically unsupported belief in vaccines as our best form of prevention against microbes. The signs are everywhere, the casualties are numerous, and denial of the evidence on the part of medical authorities is not only reprehensible but it violates their very own scientific standards. Allergies, asthma, juvenile rheumatoid arthritis, seizures and other neurologic disorders, attention deficit and hyperactivity disorder, and autism are just some of the conditions that I and many others believe to be direct consequences of an aggressive and misguided policy of over-vaccinating our children. The creation of a government sponsored vaccine injury fund to appease vaccine manufacturers and compensate the victims of this grand immunological experiment is evidence of its destructive consequences. This fund should not be construed as justification for continuing to do more of the same.

Perhaps the greatest impact of reductionism on medicine is reflected in the prevailing cultural belief that specialists know best. Both the general public and the scientific community have been duped into believing that expertise in a narrow field of knowledge involving a specified part of human anatomy or physiology will achieve the best results—better than can be achieved through a general understanding of the complex whole and the interrelationships among its parts. But history and experience indicate that greater technical knowledge does not automatically translate into wisdom and understanding. However, this should not be read as a blanket criticism of specialists whose services can be invaluable under the appropriate circumstances. Specialty medicine is best when it is utilized with full awareness of the context of a medical issue as it relates to the whole person.

Medical reductionism also provides a convenient mechanism by which responsibility and professional liability may be minimized. As long as a physician tends to the territorial obligations defined by his or her specialty, any issues that arise beyond that sphere can be considered some other specialist's responsibility. It is becoming increasingly difficult to find knowledgeable physicians willing to coordinate the overall care of patients who are often left at the mercy of a fragmented, disorganized system of independent medical specialties. The proposed solution of digitized medical records and more sophisticated computer technology is more a gimmick than anything else and should certainly not be considered a panacea for the chaos that we have come to call "managed" care.

The net result of fragmented care is exemplified by the patient who develops a string of side effects and complications while undergoing a series of medical tests, interventions, and prescriptions who, in the end, is often no better off and frequently in more compromised health than when he or she began the process. Such a chain of events may have been avoided provided there were someone willing and capable of taking responsibility for the medical whole. This is usually not the case since holism is a perspective that is underappreciated, if not absent, in most medical settings.

If a patient seeks help for an ailment and a seemingly unrelated condition begins to act up during the course of treatment, he or she will likely be referred to another specialist who deals with that particular type of problem. But when Western medicine fails to recognize the connection, for example, between headaches and hemorrhoids, simply because they are not anatomically connected, that does not constitute proof that they are unrelated conditions. From a holistic perspective, if two events occur in the same person then they must be somehow connected. Nevertheless, medical reductionism conditions us to view our symptoms as random, independent, and therefore without connection or meaning.

Medical reductionism can be conceptualized as the polar opposite of holism. One breaks things down into fragments with the expectation that one of those fragments will offer a solution relevant to a health problem, while the other assumes that the whole is greater than the sum of those fragments and that the overall well-being of the whole is the most critical factor in human health. While they are opposites in this sense, they can also complement each other. The difference is that reductionism is implied by any true form of holism—any complete holistic picture also includes a reductionistic understanding of the parts. The same cannot be said of reductionism; it is not interested in reuniting the fragments once they have been conceptually detached and isolated from the whole.

The most problematic manifestation of medical reductionism takes the form of circumstantial dissociation whereby medical events that naturally belong together are arbitrarily decoupled. It does not occur to most people to link personal medical events that have obvious temporal relationships. Specialization is so pervasive and conducive to a fragmented perspective that it makes it easy to deny connections among medical events that occur across a span of time. A child's asthma, for example, is believed to be a separate and unrelated condition that has little to do with that child's prior history of croup, bronchitis, and multiple courses of antibiotics and corticosteroids. The successful treatment of a man's arthritis with an anti-inflammatory medication is easily dismissed as irrelevant to the onset of his subsequent depression. A succession of medical events over the course of time is not considered evidence of relationship, thus making important connections easy to ignore.

We have become so programmed by reductionism that we fail to question the medical belief that a child's sudden onset of autism has nothing to do with the high fever that followed a vaccine the child received a couple days earlier. When the temporal relationship is acknowledged, experts are quick to point out that the vaccine only acted as a trigger and should not be considered the cause. Common sense should make one skeptical of such logic, but the degree to which it is accepted by society at large is a testament to the extent of its indoctrination in reductionist thinking. The possibility that vaccines might trigger autism should strike one as alarming and warrants serious investigation, not just a knee-jerk dismissal. The general public permits the medical establishment to deny such claims, largely because medicine's fragmented worldview and unbending faith in vaccines creates too much cognitive dissonance for the average person to navigate.

People are increasingly out of touch with their own first-hand experiences. They tend instead to replace them with the abstract explanations that they acquire from living in a scientifically oriented society. Many adults are incapable of adequately describing their own ailments without resorting to the use of technical medical jargon. Upon inquiring into the nature of a patient's asthma, for example, I am more likely to be given an answer that reflects medical coaching, such as "it is reactive airway disease that results in constriction of my bronchi," rather than one that indicates a more direct relationship to personal experience such as "I begin to cough and choke, sometimes to the point of vomiting, especially when I am exposed to very cold dry windy weather." We as a culture have collectively assimilated a fragmented worldview reflective of medical science's uniquely dissociated relationship to experiential reality.

It is remarkable how a science that places so much emphasis on cause and effect almost uniformly fails to identify the meaningful temporal connections between successive medical events. Reductionist medical science is in desperate need of a counterbalancing influence that acknowledges those connections, thus enabling it to put the pieces back together again. While technological medicine allows us to explore microscopic vistas never before believed possible, it lacks the moral compass and larger vision necessary to make judicious use of such valuable information. Indoctrination into the ways of Western reductionism has rendered it almost inconceivable, at least to many, that there may be other perspectives—both scientific and otherwise—capable of apprehending the larger whole and yielding better results.

Reductionist thinking allows us to view events in isolation while maintaining the illusion that it is not necessary to be cognizant of the whole in order to conduct a thorough medical evaluation. Fortunately, this is precisely what most holistic green health practitioners attempt to do. They are busy connecting the dots where conventional medicine has failed to do so. Most unconventional practitioners are generalists that view human health in its totality. They see the big picture, not just a fragmented view of a part of a part. As a consequence, they have knowledge of dynamics of disease development and healing that the medical mainstream does not recognize.

The positive value of a reductionist perspective in science and medicine is indisputable. An understanding of the parts allows doctors to set bones, deliver babies safely, and diagnose cases of appendicitis. It allows surgeons to sew up hernias, reconnect severed tendons, and remove diseased gallbladders. But reductionism ends right there. It does not concern itself with what happens to the person as a whole after the bone is set or the gallbladder is removed. And it does not allow for a deeper understanding of other factors—such as mind-body relationships or bioenergetic disturbances—that led to the problem in the first place.

When a patient reports back to his general practitioner that he experiences pain at the site of an old fracture every time it rains, and that recently the pain has spread to other joints, there is little the doctor can say regarding its connection to the original broken bone. It is simply assumed to be a new problem. As a new problem, it requires a separate medical intervention. The relationship to the old fracture is ignored and a routine prescription for a painkiller or anti-inflammatory medication is written.

The same can be said of the person who continues to complain of abdominal discomfort and digestive disturbances after her gallbladder has been removed. From a reductionist perspective, what's done is done. The remaining symptoms are seen as a different problem that needs a new solution. It would not be unusual for this patient to undergo more objective testing, which would likely turn out to be normal. For lack of any other course of action, she would then be told that there is nothing wrong and, by default, offered prescriptions for symptomatic relief.

Reductionism is so embedded in the Western worldview that we rarely stop to question its implications. In actuality, it has created a gaping hole in conventional medical theory. Medicine's philosophical orientation prevents it from acknowledging the connections between events, thus leaving a chaotic trail of medical fragments in its wake. This inability to explain the relationship between one ailment and another is precisely where holism is of most value and can serve as a powerful complement to reductionist thinking.

CHAPTER 4

Medical Mechanism

A change from the metaphor of the organism to the metaphor of the machine produced science as we know it: mechanical models of the universe were taken to represent the way the world actually worked. The movements of stars and planets were governed by impersonal mechanical principles, not by souls or spirits with their own lives and purposes.

 –Rupert Sheldrake, *Science Set Free*

Mechanistic biology grew up in opposition to vitalism. It defined itself by denying that living organisms are organized by purposive, mind-like principles, but then reinvented them in the guise of genetic programs and selfish genes.

 –Rupert Sheldrake, *Science Set Free*

A nother notable limitation of Western medicine is its decidedly mechanistic bias. The human body tends to be conceptualized as an inert assemblage of mechanical components, not unlike an automobile that periodically needs its parts repaired, removed, and/or replaced. Human life, as a consequence, becomes analogous to a machine rather than the organic whole that it is. Mechanistic thinking is closely allied with the left-brain cause-and-effect mode of perception that tends to dominate conventional medical thought. Medical events, therefore, must have clear and logically explainable physical connections in order for their relationships to be taken seriously.

Sound medical science begins with observation and description. It is an empirical process that involves observing a particular phenomenon, taking it at face value, and describing it in its purest form. Medicine's materialistic bias causes it to limit its investigations solely to physical phenomena. From these physical descriptions medical science formulates explanations about mechanisms behind disease processes, the actions of drugs, and the impact of therapeutic interventions.

Medical science allows only for causes that are tangible and quantifiable. Heartburn, therefore, is "caused" by the production of excess stomach acid, and depression is attributed to a lack of availability of the neurotransmitter serotonin. Since the conventional medical model focuses mainly on the physical it does not know what to do with a patient who claims that his heartburn is aggravated by the stress of financial insolvency or a woman who insists that her depression is a function of the unfulfilling nature of her job. Although they may evoke curiosity, such intangibles cannot be objectively measured or controlled by the tools of technological medicine. They find no place in the medical equation and are easily discarded as intriguing but inconsequential.

Although much lip service is given to the role of stress in the development of a variety of illnesses, at the end of the day, drugs and surgeries continue to be the mainstays of treatment. While medical science is comfortable translating emotions such as anger, jealousy, and fear into physiological correlates like elevated blood pressure, increased heart rate, dry mouth, and so on, it has no understanding of the direct impact of such emotions on health and illness. The relationships between anger and back pain, or grief and arthritis, for example, are not spoken of in such terms because they cannot be assigned tangible mechanisms of causation.

In similar fashion, "coincidental" events are easily dismissed as such because no mechanistic connection that explains their relationship can be identified. If a man dreams that his head exploded after flying into a rage and he subsequently experiences the onset of migraine headaches, this will not likely influence the therapeutic decisions of a conventional physician who will still tend to opt for a pharmaceutical approach. That there may be a connection between the dream and migraines should certainly prompt an inquiry into whether anger issues may be at the root of the problem. Medical practice is riddled with this type of thinking, which tends to dismiss the firsthand experiences of actual patients and their real life problems in favor of predetermined mechanistic concepts of disease etiology and pathology.

Consider for a moment a person with chronic back pain who, under the care of a holistic practitioner, suddenly experiences relief from pain that coincides with a strong emotional release, which is then followed by the onset of a head cold. This chain of events, if acknowledged at all, would likely be judged by mechanistic science to be coincidental. Each event would be assumed to be discrete, unrelated, and therefore without meaning. While a regular doctor might be tempted to offer an antidepressant and symptomatic treatment for the cold, a holistic practitioner would tend to view the cathartic relief followed by a cold as a desirable healing crisis that carries with it a positive prognosis regarding the back pain. A holistic perspective does not dismiss the possibility of meaningful connections between medical events just because discernable mechanistic relationships cannot be identified. It acknowledges the reality of additional types of relationships, including mind-body interactions.

When a prescription for arthritis of the knees brings relief but is followed by a heart attack, the link between the two is easy to ignore because it lacks a conventional cause and effect explanation. It is simply assumed that the knees have nothing to do with the heart. Many thousands of such events actually did occur and were overlooked until finally the death toll could no longer be ignored. That is why the anti-inflammatory drug, Vioxx, was taken off the market. But few seem interested in investigating why this happened because the answer lies somewhere in that forbidden frontier outside the bounds of medicine's materialistic, mechanistic, reductionist domain.

If, on the other hand, we trust our instincts and note that something is fishy when such events occur, we begin to look beyond the overly simplistic parameters of conventional cause-and-effect. Holistic theory takes it for granted that when two supposedly coincidental events take place within the same person, then they are likely to be connected, even when a mechanistic explanation is nowhere to be found. Any factor that affects any one part of the body-mind-heart-soul, by definition, impacts the whole, whether or not that impact is discernable or measurable.

The mechanistic causes preferred by modern medicine are physiological, biochemical, and immunological causes but, increasingly, genetic mechanisms have become the favored explanations. Belief in genetic determinism serves the reductionistic, materialistic, and mechanistic desire to find final causation in small physical parts. Such evidence, once discovered, it is believed, will be indisputable. If we can just unlock the genetic code and learn how to manipulate the right genes then the keys to good health will be at our fingertips. This misguided fantasy is allowed to persist only by repeatedly ignoring the warning signs and failing to make the connections regarding genetically altered life forms and the serious problems they are creating for agriculture, the environment, and microenvironment. Unwanted phenomena that threaten to derail the objectives of the biotechnological agenda are conveniently overlooked when they do not support the –isms of its predetermined worldview. Although some conditions are genetically determined to varying degrees, most illnesses do not fall into this category, regardless of medicine's tendency to assume that they are.

True to form, the reductionist search for mechanistic answers has found new life in the nascent field of epigenetics. Taking note of a phenomenon that has been recognized by some for more than a century, microbiologists and geneticists are now showing sudden interest in the notion of the heritability of certain types of traits, which were previously thought not to be heritable. The source of renewed interest is the discovery that it is possible for gene activity to be influenced through a mechanism involving epigenetic markers, which does not involve alteration of DNA gene sequencing. Genes, once thought to be fixed from birth, are now believed to be susceptible to various environmental factors, including toxins, nutritional factors, exercise, and even emotional states. The terminology is new but the concept is the same: excitement is building over the possibility of being able to intervene at the microscopic level through the manipulation of gene expression.

Charles Darwin's predecessor, Jean-Baptist Lamarck, had proposed the idea that traits acquired during one's lifetime could be passed along to offspring. Of course, that was rejected as nonsense two hundred years ago when it was determined that human traits could only be transmitted through genes, which, once inherited, remained permanently set over the course of one's lifetime. Now, thanks to epigenetics, scientists are beginning to rethink the notion of Lamarckian inheritance.

My own two hundred year old field of homeopathic medicine takes it for granted that many acquired traits are routinely passed down to offspring. It is a distinct factor that I consider in the evaluation of my patients. Many behaviors that have no logical explanation for their presence can be traced to parents who exhibit similar behaviors. For example, the intense distress of a child who comes into the world crying inconsolably may be traced to a mother's grief over the death of her parent two years prior to conception of the newborn. Or, a father who entertains suicidal thoughts after losing his job may subsequently conceive a child who begins to exhibit depressive tendencies early on in life. An expectant mother who craves salty foods may give birth to a child who craves the same. These are routine observations that all good homeopathic practitioners take note of in their encounters with patients.

The question in my mind is, why is something that has been so plainly evident for so long, at least to many alternative-minded individuals, suddenly a hot topic of scientific investigation? The idea of the heritability of acquired traits is nothing new, however, the scientific explanation is new. The answer, of course, has to do with the discovery of tangible biochemical epigenetic mechanisms that can influence gene expression, thus making it compatible with existing conventional medical theory. Never mind that the heritability of acquired traits should be evident to any keen observer; science was simply unwilling to accept such a phenomenon if it could not be explained in materialistic, reductionistic, and mechanistic terms. Now, with epigenetics, scientists can focus on epigenetic factors and remain faithful to their biases, while continuing to underestimate the role of the psyche in mind-body medicine.

There is no doubt that the science of genetics has made significant contributions to medicine. And while there clearly may be some value to the science of epigenetics, it will most likely guarantee the continued overemphasis on physical causation and the downplaying of the role of human consciousness in health and healing. The danger is that scientists will focus on epigenetics in the belief that they can alter hearts and minds by methods of genetic manipulation—and there is no telling what kind of damage may be wrought along the way.

The same issue is at play in psychiatry's biological orientation toward mental illness. Psychiatry is preoccupied with brain chemistry as a determinant of psychological states, rather than the reverse, which proposes that mental-emotional states drive brain chemistry. Biochemistry receives increasing attention while the human psyche receives less and less. In both cases, medicine fails to recognize the primacy of the psyche and the power of its impact upon physical, emotional, and mental health.

Since medical science only investigates so-called material reality, it naturally formulates its explanations in physical terms. The possibility of non-physical explanations has been excluded by the metaphysical parameters of science itself. It would violate the canon of scientific methodology to do otherwise. All disease explanations are framed in terms of the materialistic, reductionistic, and mechanistic boundaries set out by medical science. When we are fully cognizant of medicine's beliefs and biases, it becomes clear how much the various –isms overlap and serve to reinforce each other.

There should be no problem with the explanations proffered by medicine so long as they are acknowledged by its practitioners to be defined by the self-imposed boundaries of its chosen scientific methodology. Problems frequently arise, however, because medicine has been known to ignore those boundaries. It often claims to be able to explain more than is within the scope of its understanding. Although it consciously aims to study the material nature of the physical body, it occasionally and unjustifiably presumes it has the authority to weigh in on issues involving non-physical matters such as psyche, soul, emotion, thought, consciousness, energy, purpose, and meaning. At best it claims that these immaterial factors do not fall within the scope of medical concerns. A harder line contends that such factors have nothing to do with human health or that they are mere by-products of brain electrochemistry. At worst, medicine denies that they exist at all.

A particularly problematic issue arises when descriptions of observed physical phenomena and their associated details are used to justify knowledge of causation. All scientific endeavors involve phases of observation followed by proposed hypothetical explanations, but it is unacceptable when description and mechanistic speculation become substitutes for causation. Attributing causation is no simple matter and should involve a great deal of consideration before conclusions can be made. Such explanations are acceptable as reductionist explanations of the local and limited phenomena that they describe—but they do not necessarily translate into sound evidence regarding the greater issue of causation.

The relationships between most phenomena observed by medical scientists are more accurately characterized as *associations*. Western medicine arbitrarily and almost uniformly confuses association, in good faith and in bad, with causation. The vast bulk of medical science is observational and descriptive and yet it commonly conflates the observed associations between phenomena with evidence of causation. Scientists and the general public are often dazzled by such assertions of causation even though they frequently fail the test of practical application. As we shall see, most assertions of causation are really exercises in logic that serve to support medicine's preconceived notions of disease. To state the case more directly, they are rationalizations based in analytic thought but not in science. They are seductive hypotheses unsupported by evidence.

For years medical scientists have told us that depression is caused by a chemical imbalance in the brain involving the neurotransmitter, serotonin. This explanation meets the materialistic and mechanistic criteria of being a tangible cause involving a biochemical process. It is logical, sounds very scientific, and can be localized to the brain. But is it true? Rarely do we hear the equally plausible argument that states of mind can alter brain chemistry. Which came first, the altered mental state or the biochemical change in the brain? The materialist bias of medicine, of course, assumes the latter to be true.

An additional example may serve to illustrate the point regarding causation. A bystander with some knowledge of anatomy *observes* me throwing a baseball. He may *explain* that the flexing motion that I make with my arm is due to a contraction of muscles that results in the arm bending at the elbow joint. It is an entirely different matter, however, to identify what *caused* me to bend my elbow. Was it a nerve impulse from the spinal cord that caused the muscles to contract? Or was it a brain neurotransmitter that caused a nerve impulse to activate the muscles? Perhaps it was a prior thought that triggered the neurotransmitter to be released. Maybe it was my intention to throw a baseball, conceived earlier that morning, that ultimately led me to flex my arm. You can see how causation is a tricky matter that cannot be determined without thorough awareness of the situation as a whole. Attribution of conventional medical cause and effect becomes more credible when it is acknowledged that it applies only to limited circumstances and local conditions.

Some, like British Empiricist philosopher David Hume, have argued that causality itself is not a fact but rather a construction of the mind. We observe various phenomena and "cause" is the name that we give to the connections between those facts or events. But no one has ever seen a cause; we only perceive one event followed by another. We simply assume causation, just as in other cases we arbitrarily deny causation. Carl Jung's famous definition of synchronicity as "meaningful coincidence" is a concept that conventional left-brain oriented science has a great deal of trouble comprehending. The implication of synchronicity is that there is some deeper underlying *cause* for a set of phenomena that, on the surface, appears to be coincidental or causally unrelated. Since we can only have probable knowledge of patterns of events that do not have known mechanistic explanations, science tends to discard them as unworthy of further consideration.

Nevertheless, medicine is quite fond of presenting its material descriptions and explanatory logic as evidence of mechanistic causation. Likewise, it shows little interest in acausal synchronistic phenomena even when they are repeatedly encountered in clinical settings. The relevance of this philosophical discussion becomes apparent when we attempt to treat the so-called causes of a medical condition only to find that our results fall short. A simple example is a child's ear pain and inflammation that is blamed on a bacterial infection. Even if antibiotic treatment resolves the condition, what are we to conclude when the same ear becomes repeatedly infected? Most holistic practitioners argue that antibiotics encourage the recurrence of such problems, and should therefore be used sparingly. If this is the case, then how can we ascribe causation to a bacterium? And what are we to think when the same bacterium is cultured in another child's ear but that child has never had an ear infection? The problem lies in our shortsighted understanding of causation.

The concept of causation is a profound, far-reaching, and complex issue. First, it should be understood that there are degrees of causation. A *proximate* cause in medicine is one that is easiest to assign because it has an immediate and close relationship to a given ailment. It tends to be an explanation that is closest in space and time to the illness in question. It is also important to understand that all proximate causes involve arbitrarily chosen endpoints. In other words, medicine settles on a chosen causal explanation after it ceases to look for more satisfactory or comprehensive explanations. Since medicine has little interest in holistic theory, this becomes quite easy to do. However, there is nothing wrong with assigning proximate causation for an illness, as long as it is understood in proper context to be a proximate cause. This is no small point of philosophical contention. This becomes clear when treatment aimed at the proximate cause of an illness fails to get to the bottom or source of the problem.

Let's use asthma as an example. Depending on whom we ask, we may be told that asthma is caused by bronchospasm, a form of constriction of the airways of the respiratory tract. Alternatively, it may be claimed that it is due to inflammation of the airways. Some attribute it to an allergic phenomenon, focusing their attention on the environmental triggers of asthma such as dust, mold, pollen, animal dander, smoke, and so on. Still others suggest that it may have a genetic component. No doubt, these can all be factors in asthma, but it is important that they be understood as proximate causes.

Asthma treatment, accordingly, is aimed at these proximate causes. Bronchodilators are used to open constricted airways, anti-inflammatory drugs like corticosteroids are prescribed to fight inflammation, antihistamines are used to combat allergic reactions of the airways to allergens, and patients are instructed to avoid possible triggers. And when we believe asthma to be a genetic condition, we lose incentive to seek further solutions, because it conditions us to accept the idea that asthma is a problem that one must live with it for the rest of one's life.

The best way to spot a proximate cause is to ask the question, but what causes that? For instance, what causes the airways to constrict in asthma? We may be told that it is caused by an allergic reaction to a variety of allergens. This is, in essence, another proximate cause to which we may respond, why does the allergic reaction occur? And we may be told that this involves the tightening of smooth muscle in the airways. Well, what causes the smooth muscle to contract? At this point we may be offered a mechanical explanation in terms of the movements of myosin and actin filaments that comprise smooth muscle. But why do the filaments move like that? The explanations tend to become more detailed and reductionistic as we go on, often involving smaller and smaller parts and, in this case, we are likely to be told that it amounts to biochemical reactions involving proteins, enzymes, and ions. But why does that happen? The logic of proximate causation eventually leads back round again to the originally stated proximate cause—it is triggered by some type of allergen.

Note the circularity of this type of mechanistic argument. Explanations begin at a given point in a chain of events and eventually wind up back at the same arbitrary starting point. Medical researchers can spend whole careers studying one small aspect of this chain of events in the hope of making a contribution toward a cure for asthma. But very few are concerned with studying the bigger picture. This begs a number of important questions. Are these purported causes really causes? Or are they arbitrarily chosen proximate physical endpoints as defined by the mechanistic bias of Western medical thought? Are these causes at all, or are they merely detailed *descriptions* of the physiological events involved in asthma? Are they, in reality, just the phenomena *associated* with asthma as observed by medical scientists, any one of which can be arbitrarily designated as a causative factor?

It is possible to look beyond proximate causation in the hope of discovering *actual* or *ultimate* causation. Given the arbitrary nature of proximate causation, it seems reasonable that more reliable causes should be readily identifiable. But when we search for deeper causes it tends to raise even more questions. Can we even know the ultimate cause of, say, asthma? Does medicine have any interest in the ultimate cause of asthma? Can we ever know the actual cause of any illness?

Most forms of energy medicine take it for granted that illness is ultimately an energetic phenomenon. Why is medicine not interested in the bioenergetic dimension of disease? Does one's bioenergetic state have anything to do with one's physical, mental, and emotional health? Or is the cause of illness a complex, mulifactorial issue that will always elude reductionist attempts to pin it down to one or two overly simplified factors?

Medicine's preoccupation with proximate causation is a function of its desire to generalize, which is to say, to find similar causative factors for all cases of asthma. Here, again, we see how medicine imposes its perspective on the data, rather than allowing the data to guide its understanding of asthma. In its desire to find a common denominator for asthma it overlooks the reality of patients, each of whom tends to experience asthma in different ways. Careful medical history taking reveals that each case of asthma is triggered by different factors and has its origin in unique circumstances. If asked directly, some patients will be able to offer reasons for why they think they developed asthma in the first place. But medicine is not really interested in this type of information because when we look for actual causation, the answers tend to become less concrete and more intangible.

It is one thing to ask what it is that causes all or most cases of asthma, but it is a very different question to ask why a specific person has contracted asthma in the first place. If there are potentially useful ideas regarding the actual cause of a given patient's asthma, is there something that can be done in light of that information to contribute to its treatment? Questions of this nature are usually avoided by conventional medical science because they are not compatible with its reductionistic, materialistic, and mechanistic principles. It would require genuine individualization of treatment in lieu of a one-size-fits-all approach. Taking such questions seriously could throw the medical world into an existential crisis and, so, it is just as easy to ignore them.

Questions of ultimate causation inevitably take us into the metaphysical realm. There is nothing that medicine dislikes more than questions that are subject to open-ended inquiry and ongoing debate. Medical science's need for certainty runs contrary to such ambiguity. It is reluctant to concede that we may not be able to know the answers to some questions. As a consequence, it commonly fudges the answers by overconfidently presuming that non-physical considerations have nothing to do with the material problems at hand.

When pressed to account for the role of non-physical factors in the development of illness, medicine has been known to respond by casting aspersions. Patients seeking answers to their questions may be reminded of their ignorance in medical matters. Likewise, alternative healing modalities are usually dismissed as unscientific or pseudoscientific. Thus, the problem of causation is solved in an intellectually dishonest way. Medicine turns a blind eye to anomalous information that it deems too undignified for the investigations of serious science, and opts instead to focus on the strictly physical. As a result, a good number of medically relevant topics do not receive the attention that they deserve.

Proximate causation should never be mistaken for actual or ultimate causation. All conventional medical explanations involve proximate physical causes because they deny the role of the deeper underlying energetic, mental-emotional, and spiritual dimensions of illness. Medical mechanism, by definition, does not seek to look any further than the physical. It prefers localized, physical, simplistic cause-and-effect explanations based upon associations that may not be causes at all. Like reductionism, it runs in a direction opposite to holism.

While proximate causation is limited and finite, ultimate causation may not be knowable. However, this is no reason to throw up one's hands in resignation. Certainty always comes in degrees. Potential solutions should not be disregarded because they cannot be verified with absolute confidence. There is a middle ground. The spectrum of causation stretches from the more local, concrete, proximate, and reductionistic, to the most comprehensive and holistic. A severed hand tendon requires the predominantly local and proximate expertise of a hand surgeon. Most chronic diseases such as arthritis, asthma, and diabetes are better served by a physician with a broad knowledge base and a more holistic outlook. If we are willing to look, we can often find deeper causes along the spectrum of causation. When we make a concerted effort to address those deeper causes we tend to get more thorough and long-lasting results.

I have had the chance to treat more than a few hyperthyroidism cases over the course of my career. Conventional medicine, having no other option, seeks to address the physical phenomenon itself, usually by destroying the thyroid gland with radioactive isotopes of iodine or by removing the gland surgically. The assumption made is that the "diseased" thyroid must be neutralized. Treatment usually results in the reverse condition, hypothyroidism, which must then be compensated for with thyroid hormone replacement therapy.

With more options at my alternative medical disposal, I am able to explore other potential etiologies that involve more than just physical factors. As it turns out, I have traced several cases of hyperthyroidism to emotional factors involving grief. In each case, the patient had suffered a devastating loss just prior to the onset of hyperthyroidism. In each case, the patient was still struggling with the emotional distress that comes with losing a loved one. In each case, I prescribed a homeopathic medication known to be able to release a person from the grips of overwhelming grief, thus allowing emotions that were previously stuck in a negative feedback loop to be processed in a normal healthy manner. And in each case, when the person began to process those emotions, their hyperthyroid condition started to resolve, thereafter necessitating no further medical intervention.

Medical science is incredulous when it hears such stories, which, to me and other holistic practitioners, are not unusual. The inability to believe in case histories like this are a function of unconscious philosophical programming that precludes all but material causes and is skeptical of any phenomenon that cannot be assigned a tangible mechanism of causation. To me, the mechanism in each case was clear: severe grief triggered the onset of hyperthyroidism. The problem is that medicine does not really believe in something as intangible and unquantifiable as psychic causation. It is nearly impossible for medicine to conceive that a thyroid problem could have its causation in the mind, and even less believable that the solution would involve the resolution of an emotional issue. Rather than evoke curiosity, it is common for skeptics to demand tangible proof of a mind-body connection before they are willing entertain such possibilities.

As far as I am concerned, the outcomes of treatment constitute the best evidence. In each case the grief lifted and the so-called diseased thyroid returned to normal. I am not claiming that all hyperthyroid conditions are of the same nature but, for some, there is a definite connection to grief. This brings us back to my original point, the issue of causation. The most proximate causes are naturally assumed to be located in or around anatomical sites of dysfunction. In this example that would be the thyroid gland. Deeper causes, however, are not necessarily physical, and often involve the mental, emotional, spiritual, and energetic states of individual patients.

The same argument applies to a wide variety of medical phenomena.

In accordance with medicine's mechanistic and materialistic standards, it is not possible to understand why the suppression of arthritis with an anti-inflammatory drug, for example, could lead to depression. Similarly, it is not conceivable that a child's attention problems in school could be rooted in the use of a steroid inhaler prescribed for his or her asthma. And yet, in my estimation, phenomena of this nature are the rule, not the exception.

When we remove the straightjacket that limits medical thinking, we find that there are varying degrees of causation. Most are not causes at all but merely associations. Some are relatively superficial causes, others are partial causes among many factors, and a few are deeper causes that come closer to the true underlying source of illness. Medical science does not have the inclination, flexibility, or patience to search for deeper causes that may lie waiting to be discovered. It prefers predictability and certainty and underestimates the value of subjective experience, the indefinite, and the immeasurable.

Medicine will eventually have to accept the basic principle that, with certain exceptions, treatment based on proximate causes yields short-term palliative results, while holistic treatment based on a more sophisticated understanding of causation brings us closer to genuine healing. Human illness is not fixed, finite, or predictable. It will always retain a good bit of mystery. In the end, I believe that medical treatment outcomes and success rates will remain dependent upon our ability to perceive causation through the perspectives of both reductionism and holism.

CHAPTER 5

Medical Rationalism

Man is a rational being, but he is also something much more. Reason is one of his tools—not his definition.
  –Gai Eaton, *Science and the Myth of Progress*

Some folks trust to reason
Others trust to might
I don't trust to nothing
But I know it come out right
  –Robert Hunter, *Playing in the Band*

The philosophical footing for the Scientific Revolution in the Western world is often credited to the seventeenth century French philosopher, Rene Descartes. Also known as the father of modern philosophy, Descartes' writings created the framework for what has become the hallmark of modern scientific man—his thinking mode of being in the world. Contemporary rational scientific man is what he is precisely because he thinks; he values thinking above all other modes of being. This is the cultural bias of modern Western societies. We are inclined to believe that we can *think* our way into, through, and out of any situation that we set our sights on.

Descartes' historical prominence and association with modern philosophy have a great deal to do with his relationship to rationalism. Rationalism, of course, is the unspoken consensus philosophy embraced by contemporary Western cultures. It is the belief that our opinions and actions should be based, above all, upon logic and rational thought. All reliable knowledge must, similarly, derive from the human capacity to think. The corollary is that other, non-rational features of the human psyche such as emotion, intuition, dreams, meaning, moral imperatives, religious belief, and metaphysical principles are inferior to the rational function and do not constitute reliable sources of knowledge. In essence—and this is a vital point—analytic *thought* is the only thing that can provide certainty while *experience* is not to be trusted.

At its height, rationalism took one of its most extreme forms in a philosophy called *logical positivism*. Proponents of logical positivism believed that scientific methodology was the only means to obtain knowledge of the world. At the same time, they defined science in the strictest of terms. In other words, only empirical evidence gathered by observation via the senses and then subjected to the rigorous examination of logic and mathematical analysis could produce knowledge of reality. Any knowledge worth knowing should be able to withstand this type of scrutiny if it was to be believed. Anything of value had to be provable by this method of observation and analysis. Any phenomenon that could be considered factual had to first be able to be observed by the senses and then able to survive the rational scrutiny of logic. Only scientific knowledge was useful knowledge. In fact, logical positivists claimed that the only acceptable truths, other than those that could be empirically observed and verified by science, were the truths of logic and mathematics.

As a consequence, ethics, theology, metaphysics, anything of a subjective nature, and much of philosophy itself were rejected by the logical positivists as meaningless and useless. It should not be surprising, given the scientific bent of Western cultures, that this was the dominant school of philosophy through much of the twentieth century. For a time, it was the principal school of thought, especially in most academic departments of philosophy of science.

It is also no small point to make that by the 1960-70's, most academicians considered logical positivism to be dead in the water. Although rationalism is the product of a dying paradigm, it has continued to hold strong sway in the United States and Europe since the early 1900's. While most extremist versions of rationalism have been thoroughly discredited by academic philosophy, rationalism, nevertheless, continues to have a powerful influence on practicing scientists who apparently haven't gotten the memo that the old school is running on fumes. This disconnect between our evolving understanding of a holistic universe and an outdated conception of a material world that must be dissected and subjected to reductionist principles of logic is an enormous source of scientific dysfunction. This is particularly true when it comes to the practice of conventional Western medicine.

Descartes' famous statement, *cogito ergo sum*, was a declaration of the means by which he could assert with confidence the truth of his own existence. Here, in his *Principles of Philosophy*, Descartes presents his most influential discovery to the world:

> *... for there is a contradiction in conceiving that what thinks does not at the same time as it thinks, exist. And hence this conclusion, I think, therefore I am (cogito ergo sum), is the first and most certain of all that occurs to one who philosophizes in an orderly way.*[5]

Descartes believed that by virtue of the fact that he could reflect upon himself that this was evidence of the reality of his existence as an individual being. By this step of logic he established thinking as his method of discovering the truth of his existence. Descartes believed the mental faculty to be superior and placed it at the pinnacle of a hierarchy of human traits. I would contend, however, that this is one of those unexamined assumptions that are incorrectly believed to be unassailable principles that form the bedrock of scientific methodology.

Of course, a belief that the mental faculty is superior to all others is a highly prejudicial assumption to make. One could just as easily arrive at the same conclusion by asserting that "I emote, therefore I am," or "I intuit, therefore I am," or "I sense, therefore I am." In fact, emotion, intuition, and physical sensation tend to be experiences that precede cognition. Usually, it is only after one experiences something that one then has thoughts regarding that something. We reflect back upon our experiences in order to draw conclusions about them. Thought is usually a secondary process that reflects back upon primary experience. First-hand personal experience, I would argue, provides a more direct apprehension of one's being in the world.

Some Eastern philosophies would contend that Descartes' "I" is merely an illusion. It is a construction of the ego, which is invested in its own self-autonomy. "I" is an illusion that establishes me as independent and separate from the rest of existence, whereas, in an Eastern construction of reality, the true self is but a reflection of the ground of being, which dissolves into the greater whole. "I" and the whole of creation are mere reference points along the spectrum of existence. According to this perspective, my little ego, which insists upon the certainty of "I" is really a trick of the mind—the very same mind that Descartes claims is the source of the certainty of his existence.

Furthermore, many believe there is a deeper form of knowing that is independent from thought. It is a precognitive, pre-rational, even pre-experiential form of knowing that can be roughly formulated as "I am, therefore I am." The Biblical Exodus tells us that Moses inquired as to the name of God, whereupon God replied variously "I Am that I Am," "I Am," and "YWHW." The Hebrew reference for God is YHWH, the true pronunciation of which is uncertain. Other interpretations of YWHW or Yahweh also include "I Am Who I Am," "I Am He Who Is", "I Am Who Am," and "He Brings Into Existence Whatever Exists."

Most scholars would agree that Hindu culture is the repository of the world's most ancient spiritual tradition. Within that tradition, Indian Vedantic religious texts called Upanishads include several sayings that are understood to be of great significance. One such statement has been translated from Sanskrit to read *Tat tvam asi*. This statement is interpreted variously as "Thou art that," "That thou art," "That you are," and "You are that." The meaning of *Tat tvam asi* is that the deepest self, the soul, or consciousness in its purest state, is understood to be the divinity within you, and the divinity within you is continuous with, if not identical to, Ultimate Reality, the ground of all being. One does not come to know this truth through thought, analysis, or reflection. It is a mysterious and deeper form of knowing that recognizes divine reality as that which precedes the created universe and that which is the origin of all phenomena and all consciousness.

The purpose of this discussion is not to claim that I am somehow equal to a god, but to highlight the idea that it is possible to recognize or to *know*, before even thinking about it, that we are divine beings and that we are part of a greater whole. This knowledge of one's being exists before thought, before rational discourse. Like the Biblical God, I can *know*, without forethought, that I am. And like Vedantic truth, it is possible to *know* that I am not discontinuous from all that I perceive outside of me—that which conventional science defines as objective reality, the domain of its operations. Neither truth requires explanation or defense. They are simply truths that one comes to know through divine grace. They are fundamental principles that form the ground of a particular worldview, and are quite different from the principles that inform modern science.

Scientific skeptics tend to claim that religious beliefs are faith-based. In other words, such beliefs are based upon second-hand knowledge that people have learned from someone or somewhere else, and that they take this information on faith, without proof. After all, they argue, there is no proof other than the proof provided by science, which is the type of proof that comes from empirical observation and rational analysis. Even contemporary religions have been seduced by this bold claim and have fallen for the erroneous notion that there are only two antithetical options: scientific proof or religious faith. But this fails to take into account the countless individuals throughout history who have possessed a different form of first-hand knowledge—a knowledge that is not obtained through faith but is known simply because one knows it deep in one's bones.

There are a variety of types of persons to be found in the world, and I do not take issue with any particular individual's mode of perception.

Each to his own, so long as that person does not try to impose his or her worldview upon others. Some have faith, some know from experience, some know what they know, and others require logical proof. Each represents a mode of being in the world that does not have to be absolute or exclusive. It should go without saying that it is possible for me to rely on my personal experience, trust my hunches, have faith in a possible outcome, and use my power of reason all at the same time when addressing a particular situation. I do not, like Mr. Spock, rely solely on my rational function to solve a problem.

It boils down to either an arbitrary choice or a function of one's characterological makeup, akin to the differences between right and left-brain dominant individuals, when one pins the certainty of one's existence on the capacity to think. Likewise, to place reason above other human faculties of perception and modes of experience and then rely upon it alone to guide scientific inquiry is both arbitrary and unwise. Western culture has been saddled with this skewed perspective regarding reality ever since the ancient Greeks raised the rational mind to its venerated status above other human modes of experience. Orthodox medicine followed suit in giving priority to the rational functions of logic, analytic thought, and quantitative measurement. Medicine uses logic to solve problems while ignoring that which it deems not rational or, to use a more prejudicial term, irrational.

Over time, the value of these other functions has been degraded, sometimes to the point of becoming vestigial traits that are so neglected as to be unrecognizable. This is epitomized by the stereotype of the intellectual who can dazzle his audience with his encyclopedic knowledge and Socratic skills but hasn't the foggiest awareness of his own emotional life. Western culture as a whole is highly rational and, as a consequence, is becoming increasingly dissociated from emotion, intuition, and personal experience. We are increasingly susceptible to ill-conceived advice to guide our lives because we are out of touch with our inner selves. I believe this trend, among other reasons, to be a strong contributing factor to the modern rise in mental illness.

The net result of this influence is an odd standard of medical proof that requires that something be quantified and logically explained before it can be believed. Many alternative forms of healing are dismissed when they do not meet these requirements. Acupuncturists and homeopaths can point to centuries of clinical experience and millions of satisfied patients. Nevertheless, since no mechanism of action that explains such results in rational mechanistic terms to appease scientific authority can be produced, they are denounced as "pseudoscientific" therapies that are inferior to "science-based" medicine. This turns out to be a double standard, a capricious form of bias, when it is revealed that many well-established surgical and pharmaceutical protocols fail to meet the same standards of proof.

Recently, the Mayo Clinic did a review of research papers published by the *New England Journal of Medicine* over a ten-year span. It revealed that 40% of all papers investigating established medical practices had determined that the standards represented by those current practices were less effective than previous standards of medical care. These so-called "reversals" applied to medical standards across the board, calling into question industry favored drugs, procedures, diagnostic methods, and screening tests. Not only did the Mayo Clinic paper highlight how subpar standards find their way into practice without adequate proof of their effectiveness, but its authors also explained how this comes to be so:

> *Many reversals have similar narratives. Although there is a weak evidence base for some practice, it gains acceptance largely through vocal support from prominent advocates and faith that the mechanism of action is sound. Later, future trials undermine the therapy, but removing the contradicted practice often proves challenging.*[6]

Here we find acknowledgment of the prevalence of faulty logic behind established medical practices. Suspect mechanisms of action provide the rationale for practices that are later found to be less effective than previously accepted practices. The fruits of excessive rationalism are medical practices that appear at first glance to stand the test of logic but ultimately fail the test of practical science. In the words of the authors, they represent medical explanations that require "faith that the mechanism of action is sound." Such rationales require faith precisely because they have not been verified by scientific method. These practices may continue to find favor even after they have been discredited by the latest research.

One such faith-based practice cited by the authors involves the treatment of vestibular neuritis. This is a condition of the inner ear that affects the sense of balance and is presumed to be of viral origin in most cases. In spite of its supposed viral cause, physicians variously treat the condition with corticosteroids, antiviral drugs, or both. This inconsistent standard of treatment persists in contradiction to medicine's assumption of viral causation. Furthermore, the same erratic therapeutic approach continues even after the most up to date conventional research indicates that only corticosteroids are capable of helping the condition. Here we have an example where presumed causes based on logical assumptions, mechanisms of action believed to be true, actual clinical practices, and research findings all clash and contradict each another.

Modern medicine relies primarily on a rational approach to human health. It is a fundamentally left-brain mode that places a premium on logic and the quantitative. It places excessive value on lab results and the statistical abstractions of research studies while it downplays the immediacy of patient experience. Physicians are more inclined to believe their concepts, theories, and mathematical calculations than the first-hand reports of patients under their care. This is the legacy of two thousand plus years of patriarchal influence. While a rational approach may be a useful method in some hard physical sciences such as geology or computer science, it is clearly inadequate when it is the dominant or exclusive approach to healing. An overly rational worldview is a breeding ground, for example, for a mindset that can casually excuse the serious adverse reactions and even deaths caused by a new drug as the necessary "risks that come with the benefits." This is not so much a scientific judgment as much as it is as a cold utilitarian calculation made by a dissociated mind trained to ignore the reality of individual circumstance.

It is worth noting that rationalism, as far as I am concerned, cannot be equated with the exercise of reason. Rationalism is an *-ism* and, as such, is beholden to its own rationality. Rationalism employs pure logic to the exclusion of practical experiential knowledge. It is a form of ungrounded mental gymnastics. Reason, on the other hand, employs rational thought but is also reasonable in the sense that it implies the use of sound judgment. The logic of reason is fair and sensible. Reason is not purely rational because it requires the subjective employment of good sense. But the imprecise and intangible nature of a concept like "good sense" renders it unacceptable to rationalist science.

The so-called causes of many diseases are, in reality, elaborate "rationalizations" influenced by Western medicine's mechanistic, materialistic, and reductionist bias. For example, we are told that high uric acid levels cause gout, constricted airways cause asthma, and some migraines are caused by vascular imbalances. While these can be considered physiological explanations of the phenomena associated with these conditions, they are not the underlying causes. They are descriptions, not explanations of causation. Nevertheless, impressive demonstrations of medical knowledge often serve to pacify inquisitive laypersons who are temporarily satisfied with such explanations—until it is found that the treatments offered for these same conditions fail to produce the desired results.

Overreliance on the rational function also serves a protective role as a form of plausible deniability when faced with unpleasant and undesirable realities. When a child suddenly regresses into an autistic state within days after being vaccinated, the physician can refer to the statistical abstractions of the latest research, which is used to argue that there is "no conclusive evidence" linking vaccinations to autism. Of course, the same statistical tactics can be employed to support just about any claim that one wishes to defend. It is the same mindset that dismisses a patient's report of his or her own symptoms and experiences as merely "anecdotal." And it is also why patients frequently come away from a visit to the doctor feeling misunderstood.

Rationalism separates the mind from experience, as if this is a desirable thing to do, and as if it will have no appreciable impact on either function or the whole. Furthermore, medical reasoning can be likened to the process of abstraction. Medicine deals in abstract entities that have been intentionally extracted from their original concrete circumstances. Medical diagnosis and treatment begins with an actual human experience of illness and then applies numerous steps of logic until that experience has been transformed into information that is compatible with its materialistic, mechanistic, and reductionistic standards of evidence. It forces real patients and their real problems to play by its artificially constructed set of rules.

Let's take two different children with ear infections as examples. One child runs a high fever, has circumscribed red flushed cheeks, and becomes inconsolable. He screams and cries, demanding to be picked up and carried, but the minute he is picked up he demands to be put back down. He makes these contradictory demands repeatedly. He is very restless and it is obvious that something is very wrong.

A second child develops a runny nose, begins to act tired and weepy, and is found to have a temperature of 99.5. She has a mild productive cough and, in spite of encouragement, loses interest in drinking fluids. Unlike the first child, she asks to be carried around and this seems to soothe her weepy demeanor.

The children are examined by a pediatrician who determines that they both have ear infections. The criteria used in each case to make a diagnosis are the general indication of sickness combined with the discovery of a red, inflamed eardrum in the right ear of the boy and the left ear of the girl.

Right away, we can see that information that conventional medicine considers superfluous is discarded in favor of factors that conform to preconceived diagnostic criteria. In other words, the characteristics that differentiate one child from the other are of little to no value. Only the signs and symptoms that they have in common and that confirm a diagnosis of otitis media are useful. The rest may make for interesting storytelling but has little impact on the doctor's decisions. Once a diagnosis is made, the therapeutic plan of action is virtually rote, thus requiring no further thought. Antibiotics are prescribed, an antipyretic is recommended for fever and discomfort, and a follow up is scheduled.

This diagnostic and therapeutic approach involves multiple steps of logic that presuppose various assumptions, which are rarely, if ever, questioned. First, regardless of the additional symptoms exhibited by each child, an inflamed eardrum is assumed to be an indication of infection. Second, without knowing if the assumed infection is bacterial or not, an antibiotic is prescribed because it is also assumed that the best way to handle bacteria is to kill them. It is also assumed that a drug to lower fever is desirable or, at the very least, will bring comfort and cause no harm. This protocol was, for a very long time, considered to be scientific medical fact. It is now up for debate in the conventional medical world.

Recall that these ear infection examples are used to illustrate medicine's predilection for rational abstraction. Modern medicine, far from being scientific in the way that the average person thinks of science, has evolved into an elaborate network of rationalizations that increasingly has less to do with individual patients and their specific needs and more to do with meeting the requirements imposed by the –isms of modern medical theory. A common sense understanding of science assumes an observant, curious, and open mind ready to adapt to circumstances as they present themselves so that the wisest choices are made in order to achieve the best possible outcomes. But what we encounter in medicine is almost the opposite. It assumes that the facts of medicine have already been established, at least in large part, and that all that remains to do is to carry out what its protocols dictate. In other words, it tends to treat specific instances of illness as if they are all the same, as if they are all just versions of the same theoretical diagnostic stereotype.

Now, let us consider these two children from the perspective of a completely different medical paradigm. Homeopathic medicine is based upon the principle of similars. The symptoms that a given substance can cause are matched up with similar symptom patterns in sick individuals. When the two are brought together, a paradoxical but very real therapeutic effect can be observed. The symptom profiles of the two examples above represent commonly observed patterns in children with ear infections. The boy's profile matches the symptom profile of homeopathic *Chamomile*, while the girl's symptoms resemble those of a specific type of flower, which, in its homeopathic form, is called *Pulsatilla*.

Twenty-five years of experience has made it clear to me that homeopathic methodology yields excellent clinical results. But the salient point here is that successful homeopathic prescribing is highly dependent upon a *complete* symptom picture of the sick individual. A partial picture that only takes into account objective physical phenomena such as fever and an inflamed eardrum would be virtually useless to a homeopathic practitioner. All of the clues taken as a whole, including objective signs and subjective symptoms such as emotional sensitivities, level of thirst, and so on, constitute the unique way in which each individual carries his or her illness. And it is this picture as a whole that guides one to the correct homeopathic medicine.

In fact, the symptoms that most children with ear infections have in common are of lesser homeopathic value. In contrast to the information obtained from a conventional medical evaluation, the best clues to help clinch a successful homeopathic prescription are the symptoms that are most unique. It is the boy's intense, inconsolable irritability that indicates *Chamomile* and the girl's weepy, gentle demeanor that indicates *Pulsatilla*. This same information is of little value to a conventional physician. The underlying assumptions regarding health and illness are different, the diagnostic and therapeutic methods are different, but the results of homeopathic treatment are just as efficacious if not superior to conventional antibiotic therapy. Even though the methodology of homeopathy is quite scientific, it tends to be categorically rejected by medical science because its premises do not fulfill the criteria of making rational sense; rational sense, that is, from the specific perspective of the *–isms* of medicine. Its paradoxical effects do not conform to the expectations of conventional medical theory, hence, it is argued, that it just cannot be possible.

While conventional medicine seeks to abstract or compress all cases of a particular illness down to their theoretical common denominators, homeopathic medicine seeks to meet each particular case of an illness on its own terms. Regular medicine treats idealized versions of illnesses arrived at through numerous steps of logic and rational analysis. It must first convert empirical data into a format that satisfies its standards for it to be of practical value. In doing so, it imposes its worldview upon the observed data. Homeopathy, on the other hand, takes all data, objective signs, and subjective experiences of patients at face value without imposing arbitrary rules that determine what is and is not acceptable. By the way, this principle of treating each medical situation as a singularly unique event is what differentiates most holistic disciplines from Western medicine.

Furthermore, the process of abstraction that pervades conventional medical theory and practice is a key factor in patient's perceptions of the medical system as depersonalizing. Abstraction literally takes the person out of the medical equation. It de-humanizes and *object-ifies* patients who, as in the cases above, become nothing more than infected eardrums disengaged from the rest of their bodies and the rest of their lives. Medicine is thus reduced to a technical matter stripped of potential meaning and significance. The point is well made by Eric Oberheim in his *Introduction* to Paul Feyerabend's *The Tyranny of Science*. Here, he summarizes one of the critical assertions made by the iconoclastic philosopher of science:

> For the objectivity and universality of science are based on abstraction, and as such, they come at a high price. Abstraction drives a wedge between our thoughts and our experience, resulting in the degeneration of both. Theoreticians, as opposed to practitioners, tend to impose tyranny through the concepts they use, which abstract away from the subjective experiences that make life meaningful.[7]

Medicine tends to justify its many rationalizations by contending that this is the way scientific method is supposed to work. Observations are made, hypotheses are proposed, experiments are performed, information is gathered, and theories are confirmed. The problem is that many times this simply is not the case. What medicine sometimes calls theories that have been tested and verified, I call rationalizations. A large percentage of so-called medical "facts" are really just logical assumptions that have never been proven.

An example of this type of logical approach to therapeutics involves the increasing use of corticosteroids like prednisone to treat a wide array of conditions. While these drugs are known to reduce inflammation, they are also well known for the serious side effects that they can generate. For this reason, they were once prescribed with great caution, but the floodgates have opened and they are now used almost nonchalantly across the board for many conditions, both in children and adults. The reason they have become so popular is because they are capable of stopping inflammatory processes dead in their tracks—at least in the short-term.

But here is where logic fails the test: it is simply assumed that reducing inflammation by this means is a good thing. There is little thought given to the bigger picture or longer-term impact of these drugs. The sometimes miraculous results achieved with prednisone become the justification for its use, regardless of the tragedies that can follow after it has been taken. In my experience, prednisone, especially when given to older adults, is often the trigger that initiates a cascade of downwardly spiraling adverse medical events. When it does not lead to death, it often leaves its victims with irreversible and debilitating chronic diseases. Nevertheless, unsound short-term logic prevails; corticosteroids are used because, in the short run, they can thwart inflammatory conditions like asthma, tendinitis, arthritis, dermatitis, and so on.

The adulation of rationality dominates medical science to such a degree that the mere use of logic, regardless of the soundness of the argument made, confers an undeserved element of truth to claims being made, simply because they are presented in logical terms. In this way, hypotheses take on an air of factual truth because they represent arguments that sound logical or make logical "sense." "Sounds logical to me" is a popular expression that may serve to convey my point here. Blinded by the erroneous assumption that rational thought provides a superior path to knowledge, somewhere along the way medical science apparently forgot the need to verify its logical but untested assertions.

Let us now examine an issue that most physicians believe to be settled medical fact. Some time ago it was determined that high cholesterol was related to heart attacks. After all, coronary arteries in heart attack victims were found to be clogged with plaques composed of cholesterol. A logical assumption was then made that cholesterol is, therefore, a major cause of heart attacks. Mistaking potential association with causation, the medical industry proceeded to manufacture and market a long string of cholesterol-lowering drugs. It was logically assumed that these drugs would reduce heart disease and heart attacks. This was assumed to be an effective strategy in spite of a long list of rather serious side effects associated with these drugs. In this manner, a chain of logical assumptions took on a life of its own. By virtue of its logical appeal, this belief regarding cholesterol has become a culturally accepted medical truth.

While studies show that cholesterol drugs can alter lipid profiles, no study has conclusively determined that this strategy reduces heart disease or heart attacks. It is simply assumed to be the case because it is grounded in logic. Now, there are a variety of alternate theories regarding cholesterol, one of which proposes that cholesterol is a vital ingredient needed for the repair of damaged blood vessels—hence, the finding of arteries caked with cholesterol in individuals with heart disease. Damage incurred through stress, the wear and tear of age, and calcification is mitigated by a self-healing mechanism employing cholesterol. This represents an equally attractive but similarly unsubstantiated hypothesis that has yet to be proven.

My point here is that both interpretations are worthy hypotheses, primarily because their logic is plausible. But logic does not magically make a conclusion true and, although the medical world would not disagree, its actions, nevertheless, frequently violate this principle. Con-

85

versely, just because something is not provable by logic or verifiable by science does not mean that it is not possible, true, or factual. As far as I am concerned, the fact that I dreamt of the ocean last night is a factual certainty that medicine will never be able to prove or disprove.

Worse yet, medicine sometimes deviates from its ostensibly rigorous standards of evidence by dabbling in far-flung flights of rational fancy. And the medical establishment seems to tolerate this, as long as it comes from within its own ranks. The relationship of salt to high blood pressure and heart attacks is an example of how rational assumptions trump actual evidence. Logic says that salt causes water retention, which increases fluid volume, which, in turn, can cause high blood pressure. While, in some cases, salt reduction may improve hypertension, it is also true that high salt intake does not necessarily cause high blood pressure. There is also little correlation between salt reduction and decreased incidence of heart attacks. These ideas regarding salt are believed to be true simply because they represent plausible logic, regardless of the actual evidence. Much of modern medicine has deteriorated to this point, whereby logic is mistaken for fact and any statement that sounds rational or plausible, as long as it is framed in materialist, reductionist, mechanistic terms, becomes a potential candidate for factual medical truth.

After decades of peer pressure and groupthink, we have arrived at this state of affairs because few have stopped to question a long standing tradition of assumptions that are believed to have been corroborated by great thinkers, philosophers, and scientists down through the ages. It is a uniquely Western bias to believe that rational thought in the form of logical scrutiny and mathematical analysis is the most reliable way, if not the only way, to obtain knowledge of certainty. Rationalism is one of the more flagrantly abused *–isms* of Western medicine. By virtue of its single-minded allegiance to the rational principle, medicine has paradoxically become more of an ideology than a science.

Rational thinking is not the only approach to the practice of medicine. Used alone, it often leads to erroneous conclusions that defy common sense. A more well rounded perspective is needed. Achieving that goal begins with an examination of our basic assumptions and most fundamental philosophical beliefs. We must be willing to admit that medical science may not know all that it claims to know, and that there may be other sources of information and knowledge that can contribute to a more thorough understanding of human health and illness.

CHAPTER 6

Medical Objectivism

The common division of the world into subject and object, inner world and outer world, body and soul is no longer adequate.
–Werner Heisenberg, *The Physicist's Conception of Nature*

While medical science's stated allegiance to rationalism is somewhat less evident, its proud claims to objectivity are particularly well documented. The American educational system gives the average student a clear impression that *science* and *objectivity* are virtually synonymous terms. In fact, we can extend the comparison to include *scientific*, *objective*, *factual*, and *truthful*, all of which are freely and commonly interchanged because they are believed by many to be expressions of similar meaning. The reality, however, is that medical science is riddled with unscientific bias, beginning with the critically important *-isms* that we have been outlining thus far.

Upon close examination, we find that there are several ways in which the claim to objectivity influences medical thinking. The following list summarizes the different strains of medical objectivism:

Although these meanings differ in many ways, they are nevertheless interrelated, especially in the context of a conventional medical approach to human illness. Let us examine these meanings one item at a time.

1. Objectivity as in not subjective
2. Objectivity as in concerned with detectable, quantifiable phenomena
3. Objectivity as in fair and unbiased
4. Objectivity as in objectifying processes and concepts as concrete things
5. Objectivity as equivalent to factual truth and reality
6. Objectivity as clinical detachment from patient experience
7. Objectivity as in value-free or value-neutral

1. Objectivity as in not subjective

Perhaps the most widely accepted understanding of objectivity comes from its ostensibly diametrical relationship to subjectivity. Objectivity and subjectivity are seen as opposites—if a particular bit of data or information is objective then it cannot, at the same time, be subjective. Of course, this is an oversimplification that fails to take into account that, in a holistic world, objective and subjective are just two sides of the same coin. However, reductionist thinking enables one to separate the two as if they are completely different things, and as if uncoupling them will have no appreciable impact upon the whole.

Objective information is information that is not a matter of opinion. Facts and events that are objective are believed to be free from subjective influence. Objective matters are thought to be matters that most rational, educated people can agree on. Objective information is the known information that forms the body of observable, verifiable knowledge. In this sense, objective information is public information that can be corroborated by others. Objective information is also tangible information that involves things like physical objects and observable events. Many of these things are quantifiable by the methods of science and mathematics.

By contrast, subjective information carries the connotation that it is lacking in objectivity. Subjective matters are personal or private matters that cannot be detected by outside observers. Subjective information, it is believed, is influenced by internal factors such as emotion, intuition, and opinion. As such, it is tainted and cannot, therefore, be objective. Objective information is believed to be free from such personal considerations.

Subjective information is thought to come from a particular person's point of view, while objectivity characterizes that which is not dependent upon just one person. Subjective matters are internal matters and are believed to be products of the mind, whereas objective information is understood to be external information, which comes from "out there" and exists independently from the mind. For example, we may both agree that there are six chairs around the table, but we may disagree as to whether those chairs are comfortable to sit in.

I would contend, however, that objective data is also a product of mind—especially the mind as defined by the particular point of view of rational, materialist science. We see here the dualistic influence of reductionism, which finds no problem with separating internal from external even though they are two aspects of the same whole. This becomes a particularly relevant point when dealing with matters of human health and illness as opposed to tables and chairs.

Subjective knowledge requires making judgment calls while objective knowledge simply "is" factual knowledge. The veracity of objective knowledge is supposedly not subject to the mind. We as a culture have a strong tendency to equate objectivity with terms like *disinterested*, *dispassionate*, and *detached*. These are the preferred terms because they imply an absence of emotional attachment, moral consideration, or subjective influence. In this sense, objective information is believed to be pure and unadulterated.

The bottom line, and perhaps unconscious motivating factor here, is that science in general, and medical science in particular, prefer to limit the scope of their activities to that which is amenable to an analytical methodology. Once something as messy, unruly, and intangible as subjectivity is allowed to enter the mix, the left brain begins to lose its grip on the governing of scientific affairs. Subjective information cannot be supported by rational thought, or so it is believed. More important, since subjectivity does not act in accordance with the rules of logic, it cannot be refuted by the rules of logic! And if subjectivity will not bend to the scrutiny of logic, then what is science to do?

There is nothing that science abhors more than imprecision, ambiguity, and uncertainty. If science were to acknowledge the significant roles played in health and illness by psyche, spirit, and consciousness, rational thought might lose its position of distinction in the hierarchy of human faculties. In this sense, medical science can be likened to a person with obsessive-compulsive tendencies who, in a losing battle, attempts to maintain order in the face of an inherently mysterious and non-linear world of unpredictable events.

Not surprisingly, these stereotypes regarding objectivity and subjectivity are built right into the use of the English language. While it is considered desirable to strive to be objective, the word *subjective* is commonly used as a criticism. Objective standards are good, while subjective judgments are suspect. Dictionary definitions often imply these skewed meanings and sometimes state them outright.

2. Objectivity as in concerned with detectable, quantifiable phenomena

Of course, the real reason that science wishes to exclude subjective phenomena from its investigations is that its methods and instruments cannot detect such information. Since subjective phenomena cannot be confirmed by objective means, they are denied a place at the table. It is hardly surprising that the rules of the scientific game have been formulated in such a way as to render subjectivity irrelevant. Subjectivity is not just neglected; it is treated as if its exclusion has no appreciable impact upon scientific calculations. By some scientific standards, subjective phenomena are considered mere by-products of the brain that do not really exist.

Unless you are a radical materialist, the notion that subjective phenomena do not exist is a rather extreme position to take. Even when it is admitted that subjective phenomena do exist, it is easier to believe that they are of peripheral importance than to amend scientific theory and methodology to accommodate them. From a holistic perspective, this turns out to be a grave mistake, especially when it comes to human medicine and the study of health and illness.

Biological science's perspective on vision and the human eye is a good example of how objective information is highlighted while subjective factors are minimized. Material considerations involving sight such as the retina, optic nerve, and visual cortex of the brain are the preferred topics of investigation, precisely because they can be located, identified, measured, and manipulated. But medical science has little to say about the subjective dimension of sight itself.

The perception of light and color, blindness, blurred vision, auras before headaches, hallucinations, and images produced by visualization techniques, are just a few such visual phenomena that are overlooked simply because they are subjective. In other words, the confirmation of their existence is completely dependent upon the person who is experiencing them. In an admirable feat of sleight of hand, medicine changes the subject, redirecting our attention to the material aspects of sight. In lieu of talking about the primary visual experiential phenomena themselves, medical science chooses to emphasize the physical structures involved in sight, like pupillary reflexes and rods and cones.

In more practical terms, unless a patient brings his or her headaches to the attention of a physician, it would never occur to that physician to investigate any further. After all, physical examination and diagnostic testing cannot detect headaches. Nevertheless, it is the subjective experience of the patient that ultimately leads to subsequent actions by the physician. Sometimes, if the headaches are persistent or severe enough, the potential tangible, detectable, physical reasons for such headaches, such as a brain tumor or aneurysm, must be investigated. Once objective causes are eliminated, it becomes a matter of trust for a physician to accept that the patient's subjective experience of headaches is "real."

When faced with a patient whose headaches are intractably resistant to drug therapy, it is not unusual for a physician to throw up his or her hands in resignation and refer the afflicted person to a therapist, in essence, implying that the headaches may be a figment of the patient's imagination. But knowing that the majority of headaches are stress-related does not mean that they are not real. The odds of a patient faking headaches are relatively slim and, so, to suggest that they are not real is to risk traumatizing an already traumatized person. Before a medical work up has begun, the patient's headaches are accepted at face value to be real but, when they fail to respond to treatment, they may be rejected as the subjective imaginings of a sick mind.

A variation on this theme is when a patient suggests to a doctor that the headaches may be caused by a medication that has been prescribed for some other condition. After consulting the literature, the doctor announces that this is not possible because it is not a documented side effect. In this instance, the physician chooses to believe the purportedly objective data represented by the literature over the patient's first-hand experience of his or her own symptoms. It can be even more disheartening when such a patient is told that there is "nothing wrong" because the diagnostic workup has revealed that everything is "within normal limits." This becomes a potentially convenient out for the physician who is constrained by the limited therapeutic options made available by his or her medical training, which, in turn, is heavily steeped in the doctrinaire -isms that we have been discussing thus far. Only by independently educating him or herself does a physician become aware of other less conventional methods of treating headaches.

There is an enormous literature that has been written on the topic of pain. It is a highly complex phenomenon that has always perplexed medical science. Pain itself is an entirely subjective phenomenon, undetectable by the instruments of science. It is sometimes possible to localize a body part where pain is experienced, but it is not possible to isolate pain itself in the same fashion that medicine isolates viruses or antigens. Unlike inflammation, which can sometimes be detected by its redness, heat, and swelling, pain leaves no such indictors in its place. There are literally no objective markers for pain—and yet there are doctors who specialize in pain and pain clinics for those suffering from chronic pain. Interestingly, medicine relies on numerical pain scales to assess and gauge the progress of patients. While assigning numerical values may confer an aura of objectivity to pain, those values are purely subjective and highly dependent upon the assessments of individual patients.

In the end, pain is an individual experiential phenomenon that cannot be pinned down by any objective, quantifiable means. For some reason, even though pain is subjective, it is taken for granted to be real, and seems to be exempt from the criticisms of scientific skeptics. Perhaps this is because of the ubiquitous nature of pain; it has been experienced at one time or another by literally everyone. If it were not for the subjective experience of pain, very few people would ever seek help from the medical profession.

By excluding subjective criteria from the medical decision making process the profession removes the human element from human medicine. This arbitrary and artificial de-emphasis of patient experience is the very thing that allows medical science to proclaim its solidarity with objectivity. Medical science has dedicated its efforts to treating detectable physical disease. As a consequence, in light of the fact that most human suffering is subjective and intangible in nature, many patients fall through the cracks of a medical system that has deliberately chosen to ignore this fundamental reality. It is no small point to make that the medical system also tends to turn a blind eye to some alternative modalities that are better suited to addressing the needs of patients on both objective and experiential levels.

3. Objectivity as in fair and unbiased

Once the artificial divide between objective and subjective had been established, it became a simple matter of promoting the one as superior to the other. Western societies have been conditioned not to question characterizations of objectivity as equivalent to being unbiased, unprejudiced, neutral, or fair-minded. The clear implication is that subjectivity does not embody these qualities. Objectivity is impartial but subjectivity is biased. Objectivity is open-minded while subjectivity is potentially discriminatory.

Objectivity and subjectivity carry these connotations because of the way that they have been historically defined by science and philosophy.

In actuality, neither should be viewed as inherently biased or unbiased. They are simply different modes of perception and being in the world. To label one or the other as partial or impartial is to attach ones values and worldview to the term, and this, in itself, is a form of bias. It may well be appropriate to exclude subjective phenomena from a system of medicine that is defined by its reliance upon so-called objective data, but it still represents a form of bias.

We are so thoroughly influenced by this built-in prejudice that many view *subjectivity* as a negative quality. Subjectivity is influenced by personal opinion and we certainly don't want to let something as unreliable as our own judgments affect our assessment of things, do we? Stated this way, the whole argument should strike one as rather silly. It is as if medical science would prefer impersonal robot-like individuals with detached computer-like minds to guide us through the health care decision-making process. Unfortunately, this is the type of mindset that the medical education system selects for through its multiple choice endurance testing and its emphasis on science-based prerequisite coursework. Individuals who meet these criteria are naturally more compatible with medicine's inherent favoritism toward objectivity.

One particularly specious argument claims that people cannot be trusted when it comes to assessing their own health status. They are not objective enough. This seems a rather odd and adversarial position for a physician to take in regards to his or her patients. Are we really that cynical that we should trust no one and assume that most patient reports are biased and that truth can only be discerned by the scrutiny of an objective-minded scientist?

We could just as easily begin with the assumption that most patient reports are reliable. Call me naïve, but I prefer to give most of my patients the benefit of the doubt before I jump to the conclusion that they are untrustworthy individuals with ulterior motives. Even if I suspected someone to be telling me lies, in the medical paradigm that I operate from, that would be considered pertinent information that could factor into the treatment that I recommend. It really comes down to a matter of common sense, the one ingredient that tends to be in short supply when we fail to take subjective information seriously.

A good example of the unfortunate consequences of such skewed thinking is illustrated by the way in which the word *anecdote* is bandied about in scientific circles. A term usually meant to refer to a brief story or description of an actual thing or event has been corrupted into a pejorative rejection of anything that cannot be backed up by hard data. Thus, case histories as reported by physicians, traditionally the mainstay of medical knowledge, are now viewed by ardent scientific skeptics as unreliable hearsay. They are accused of being merely anecdotal. The term *anecdotal* has become a virtual swear word in medical discourse. As a result, physicians have been duped and intimidated into believing that their own clinical experiences count for little when compared to the analytical data of research studies.

Do we also so distrust physicians, in similar manner as with patients, as to be unable to report their experiences in a clear and objective manner? Strangely, only after we have taken a large number of similarly anecdotal case histories and run them through the filter of statistical analysis do they then become magically transformed into acceptably reliable data. This unfortunate state of affairs has reached its current extreme under the dubious rubric of safeguarding the purity of scientific objectivity.

If we take objectivity to mean something that is minimally influenced by personal preference or bias then, yes, we might consider medical methodology to be justifiably objective in this very limited sense. But, as we have seen, objectivity in science has come to mean much more than just freedom from bias. Furthermore, this type of objectivity is, in actuality, not objective because it is founded on biased assumptions that predefine subjective human experience as either unreliable or irrelevant. Medical science's notion of objectivity is anything but objective. Clinical psychologist, Don Salmon, notes how the historical quest for objectivity, once an understandable attempt to reduce bias, eventually morphed into a much broader agenda that now influences all aspects of Western culture:

> *Over the course of several centuries, this capacity for objectivity be-*
> *came more widespread, gradually becoming a predominant shaping*
> *force in society. At first, the new attitude of perspective-taking did*
> *not involve a denial of subjectivity, only an attempt to minimize its*
> *distorting influence. By the nineteenth century, science had gained*
> *a great deal of prestige due to the enormous success and pervasive*
> *influence of technology. As a result, the objective approach that had*
> *been so successful in the sciences was gradually applied to virtually*
> *all spheres of life.*[8]

Some modern-day academic philosophers of science fall into the objectivist camp. They understandably wish to believe that there is the possibility of an objective position that transcends the individual variations of subjective experience. Alas, they, too, have taken the bait and strive to possess that illusory sense of certainty that science would have us believe is attainable. I suppose it is only natural that philosophers would aspire to achieve the same privileged cultural status enjoyed by twenty-first century scientists.

4. Objectivity as in objectifying processes and concepts as concrete things

When most people reflect upon objectivity in science they naturally tend to think of its relationship to bias. But there is another very important form of objectivity that plays a crucial role, especially in medical science, and that is the tendency to objectify. To *object-ify* something is to turn it into a concrete object. Objectification involves the translation of complex processes and patterns of phenomena into oversimplified terms that approximate a concrete "thing."

Perhaps the most well known example involves the objectification of women by men. In such cases, living, breathing, intelligent beings with ideas, thoughts, opinions, emotions, and much more, are reduced and degraded to the status of mere objects when exaggerated and exclusive emphasis is placed upon their physical attributes. It is not hard to see how prostitution and pornography become extensions of this tendency to view women as objects to be used for the gratification of men. Consequently, objectification can have real-life implications. This is only an example, and there is no doubt that the reverse occurs too. In either case, the danger occurs when the objectified version of the person is taken to be the true version while awareness and consideration for the actual person becomes lost in the process of objectification.

An example of scientific objectification would be our stereotyped understanding of the building blocks of so-called particulate matter. The old Newtonian chemical concept of an atom as a small particle of matter, a material object floating around in space, has been gradually replaced by a much more complex understanding based in quantum physics. We now conceptualize an atom as a locus of concentrated energy, and the "locations" of a variety of "particles" that form its contents are viewed as nothing more than mathematical probabilities.

The old grade school model of a molecule as an arrangement of solid atomic balls connected by bonds represented by sticks has become a gross oversimplification if not outright untruth when seen in the light of higher-level physics. Of course, this is understandable given that older atomic theory was formulated without knowledge of quantum physics. But it is not acceptable when medicine continues to do business from a similarly antiquated perspective, without taking into account the paradigmatic advances made by both physics and holistic medical therapies.

Another term that aptly expresses this tendency to objectify is *reification*. When we *reify*, we take something abstract and make it seem more concrete or more real. Scientists have a distinct tendency to reify, which is to say that they are biased toward making "things" out of patterns and processes. Science would rather deal with static nouns than active verbs because they are more easily handled and, so, it prefers to interpret phenomena through the lens of reification.

Let us return for a moment to our example of the sensory phenomenon of vision. By downplaying the subjective experience of seeing, biological science tries to redefine the visual sense in objective terms. Sight to a scientist thus becomes a matter of light striking rods and cones, which, in turn stimulates activity along the optic nerve and in the visual cortex. This is very different from the visual sense as described by a photographer, a poet, or an artist.

Similarly, the human proprioceptive sense of balance has been reified by science as a function of the brain's interpretation of the positions of microcrystals in the inner ear. The orientation of microcrystals can, for example, be related to the vertigo that a person experiences when turning in bed. By contrast, the individual experiencing a distorted sense of balance is likely to describe it as a feeling of floating, or a sensation as if the room is spinning, or as if he or she is falling to the left or to the right. An examination of the physiological status of an individual experiencing such sensations is not capable of identifying the specific nature of those sensations of disequilibrium. They are subjective sensations that cannot be accurately defined in objective terms. In other words, there will always be a subjective aspect to human experience that science cannot account for.

The tendency to reify or objectify becomes more problematic when medical science knowingly participates in the process without regard for the underlying realities, which are overlooked in the name of imposing order and uniformity upon a diverse and complex world of human illness. Something that, in actuality, is poorly understood by medical experts is stripped of its subjective elements and transformed into a much more manageable matter of objective "fact." It is easier for the scientific mind to grasp and manipulate something when it takes the shape and form of an object. This is why medicine embraces suspect diagnostic "entities" like premenstrual dysphoric disorder, erectile dysfunction, and fibromyalgia. The mere act of labeling tends to reify medical conditions. Once a condition has a name it tends to be perceived as an objective factual thing with predictable characteristics.

For example, although there are numerous variations of irritable bowel syndrome—as can be easily confirmed by carefully interviewing just a few individuals with the diagnosis—the official naming of the syndrome gives the impression that it is a monolithic "thing" with little variation. The net effect is that many who suffer from irritable bowel syndrome mistakenly assume that others with the same diagnosis have similar symptomatic experiences.

The point here is not to imply that these types of health problems do not exist, but that the oversimplified diagnostic objectification of such diverse and individual ailments, which are, in reality, complex multivariate processes, does serious injustice to the real people who suffer from these maladies. This brings us back to the cookie-cutter, one-size-fits-all, conceptualization of illness as described in my previous book, *Green Medicine.* In order to objectify disease in this manner, a significant number of distortions must be introduced into the mix, and such distortions can represent significant deviations from holistic reality.

By holistic reality I mean a genuinely comprehensive understanding of the unique ways in which a particular illness manifests in a specific individual. An accurate description of an actual illness as experienced by a specific person can only be ascertained by careful interviewing and examination of that person. This includes the personal, historical, familial, social, and environmental context within which the illness is taking place, and that illness' relationship to all other symptoms and complaints of the person in question. It requires that stereotypes and generalizations regarding the supposed diagnosis be put aside so as not to prejudice the physician.

The bottom line is that when treatment is based on a stereotype of an illness it frequently falls short of the mark. We wind up treating a reified version of the illness rather than the actual condition, and it is not hard to see why such therapeutic strategies would fail to generate successful outcomes. Given the fact that conventional medicine often deliberately ignores the personal, subjective dimension of disease, it makes one wonder how genuine, long-term patient satisfaction can be achieved.

Another important effect of objectification is that it lends credence to the illusion of the independent existence of objectified entities. It externalizes a complex phenomenon to the point that it is believed to be some thing that has a life of its own "out there," independent from the person, and independent from the mind. There is no such thing as restless leg syndrome without the person who experiences it. There is also no such thing as seasonal affective disorder—until it is conceptualized as such by the mind of medical science. It is not something out there that comes along and seizes hold of the person. Consequently, it is not something that can be eradicated or banished from the person in the sense that it can be cast out and exiled to the place from whence it came. That would require that it be an external object.

Of course, the prime example of the objectification of illness is exemplified by modern medicine's war on germs. Infectious illnesses with their diverse symptomatic presentations are thus reduced to a matter of isolating, identifying, and killing the microbes that are supposedly causing such illnesses. The shortsightedness of this strategy should be apparent in the increasingly poor results that it generates—an outcome that turns out to be a function of growing resistance to medicine's antimicrobial weapons. I discuss this issue in more depth in *Green Medicine* but, for my purposes here, I would like to draw attention to the reductionist, materialist bias in focusing all of the blame on the tiny physical entities that can be detected by the tools of medical science and destroyed by its chemical weapons. All other factors including the role of susceptibility, health of the host, and environmental factors, take a distant back seat to hunting down offending agents. A very complex medical issue is thereby oversimplified, externalized, and objectified as the germs that Western cultures tremble in fear over.

Much of modern science, as it is currently conceived and practiced, involves the objects that scientists "create" through the theories and concepts that they employ. Science's fondness for measurement and quantification only adds to the illusion of scientific objects as acceptable representations of more sophisticated underlying realities. In actuality, all illnesses are processes that take place over time, waxing and waning within living, breathing, multidimensional human beings. More than just a physical body, each human being is also endowed with heart, mind, and soul. Medical science, nevertheless, prejudicially selects certain predefined physical criteria from the greater whole in order to objectify those criteria, thus molding them into the desired object—a stereotyped diagnostic disease entity. The end result is an objectified construction that is also an abstraction from holistic reality. Disease thus becomes simplified along the same lines as an antiquated notion of atoms made up of solid particles in space.

Furthermore, when medicine excludes subjective details in the name of maintaining its objectivity it, of necessity, removes all unique and personal factors that would allow us to identify the suffering person in question. People with real and complex problems are thus reduced to the pertinent medical facts, lab values, and diagnostic images. Conventional medical methodology may therefore unwittingly contribute to the exacerbation of illness by *object-ifying*, *de-human-izing*, and *de-personalizing* the actual individuals who are in need of healing. It should come as no surprise that whenever persons are treated like things there will likely be undesirable repercussions.

What I am suggesting here is not simply that physicians personalize their interactions with patients by improving their bedside manner—although there is no doubt that thoughtfulness and kindness can go a long way toward assisting the healing process. I am recommending that the unexamined assumptions behind medical theory be reframed to such an extent as to allow the admission of subjective human experience back into medical education and practice. This will require that scientific methodology itself be altered. One might wonder how this can be done. There are a number of well-tested models employed by a variety of holistic medical disciplines that have been doing this for a very long time. And there is no doubt that the accumulated experience of these healing modalities can help point the way.

5. Objectivity as equivalent to factual truth and reality

The net effect of reification is that medical constructs created by the analytical mind begin to take on lives of their own, as if they really exist as external objective entities. The Western mind is much more inclined to believe in the factual truth of such entities. This, in turn, helps reinforce the illusion that physicians can treat diagnostic entities in isolation, separate from the rest of the afflicted person, without consequence to the greater whole.

One particularly influential trend in contemporary medical thought is the notion that even the mind is a mere subjective fantasy. There are some who believe that the brain is the one true objective reality underlying cognition. It is the brain that produces the illusion of mind through tangible physiological processes involving neurons and neurotransmitters. Neuroscience tries to convince itself of this by arbitrarily choosing to study anatomy and biochemistry while, at the same time, failing to take extrasensory phenomena like thought, emotion, consciousness, and spirit seriously.

Mind and consciousness become objectified when scientists reduce them to the anatomical and electro-physiological details revealed by brain studies. It is an easy step of logic, then, for modern medicine to equate most, if not all, mental illness with abnormalities of brain function and chemistry. It suits the medical agenda to remove the intangible aspects of human thought, emotion, and behavior from diagnostic and therapeutic decision-making.

Again, this becomes a convenient excuse when patients fail to respond to medical interventions that address only the physical body. The patient is left without options when it is suggested that the problem is not real or just the product of an active imagination. So pervasive is this mindset that some patients actually begin to doubt themselves and the realities of their own experiences. Physicist and Buddhist monk, Allan Wallace, laments this state of affairs:

> *Modern science began, with the Copernican Revolution, by displacing humanity from the center of the natural world, but scientific materialism has gone to the extreme of denying human subjectivity any place at all in the natural world. This dogma would rather deny the existence of introspection, or at least marginalize its significance, than acknowledge that, four hundred years after the Scientific Revolution, we still have no scientific means of exploring consciousness directly. In this regard, we are right now in a dark age...*[9]

There are certainly those within the medical field who do not take such a hard stance and are inclined to acknowledge the reality of mind. But even they can be intimidated by the prevailing medical groupthink and are not likely to publicly voice the opinion that the study of consciousness should be included alongside study of the brain. After all, so the argument goes, mental contents are merely subjective phenomena and, as such, do not fall into the acceptable sphere of scientific study. The result is that most medical institutions will continue to focus either on brain electrophysiology or the anatomical brain, in isolation, like an inert slab of pickled beef in a Petri dish.

One additional connotation of objectivity in medicine has to do with the determination of fact. I have already alluded to the notion that objective information is sometimes believed to be synonymous with factual reality. So the logic goes that data that has been gathered without bias, as defined by modern science, is as close to reality as one can get. Of course, this fails to take into account the *a priori* bias built into the very foundation of scientific method, which is the assumption that only those phenomena considered by medical science to be real—matter and energy—are qualified to represent reality. Objective reality is equated with reality in general. Note the self-fulfilling, circular logic that enables one to draw such a conclusion.

Technically speaking, although they are related concepts, objectivism and objectivity are not the same thing. In philosophy, *objectivity* is an epistemological term that concerns itself with knowledge and what constitutes valid knowledge. Most of what we have been discussing in this book is the epistemology of what qualifies as acceptable factual knowledge, albeit according to the biased assumptions made by medical science.

Objectivism is another name for philosophical realism, an ontological school of thought that concerns itself with the nature of reality. Ontology is a branch of metaphysics that asks questions about the nature of being. It concerns itself with existence and the delineation of what reality "is." Scientific or medical realism is deeply informed by objectivism, and is a vein of thinking that asserts that there is an actual objective reality that exists independently from our minds. This reality is the reality defined by objectivist science. Mind, by this standard of objectivism, is not considered to be real. And reality, according to the objectivist school of science, consists of the realm of objects and facts as defined by the many *–isms* that we have been discussing. The reality of scientific objectivism is determined by the facts as laid out by reductionism, mechanism, rationalism, and so on.

Another more extreme version of objectivism is committed to the notion that there is only one correct and acceptable understanding of reality. When applied to science, only matter and the conventional concept of energy are real. All else is illusion, a figment of the imagination. It should be clear, that in order to subscribe to such a viewpoint, all subjective aspects of an individual's own personal reality would, by definition, be invalidated.

The abuses that derive from the belief in science as a means of ascertaining truth or discerning the ultimate nature of reality is a central theme of this book. To claim knowledge of reality by excluding the subjective yin while taking into account only the objective yang is akin to the proverbial Hindu story of the blind man attempting to know an elephant in its entirety by feeling only the contours of its trunk. Since science is so poorly equipped to define truth and reality, it should not attempt to do so. That is a job for philosophers, metaphysicians, and theologians. Science is a tool that provides knowledge of the material world and aids in the achievement of practical outcomes, but has little to say about the nature of reality as a whole.

Although most scientists do not subscribe to such extreme viewpoints, for all practical purposes science itself, nevertheless, operates from an objectivist perspective. It functions as if the only valid truths are the objective facts delineated by science. All other subjective phenomena are not pertinent to its goals, if not completely non-existent. For this reason, a clear and fully conscious understanding of these conceptual dynamics is crucial to the way science and medical science are conceived and practiced. Here, Alan Wallace alludes to this disconnect between scientific theory and its actual practice:

> *Scientists have not proved the hypothesis that no truths lie beyond the domain of science, nor have they confirmed the hypothesis that no methodologies other than those of science can expand the horizons of human knowledge. But, with a leap of faith, scientific materialism accepts both those hypotheses as if they were established facts.*[10]

This is not just philosophical speculation intended for the satisfaction of high-minded thinkers. It portends real life consequences perhaps more than any other line of contemporary philosophical inquiry. The *-isms* of science have poisoned our thinking and stifled our ability to question the materialist reality that we either consciously or subconsciously take for granted. This book, at bottom, is a call for a return to open minded inquiry, free from the tyranny of mainstream medical intellectual constraints.

As far as I am concerned, my thought of an ice cream cone, my dream of a dog chasing me, my sensation of burning in my chest, my guilt over not having apologized for something hurtful I said, and my intuitive sense that today is a good day to contact an old friend, are all subjective "facts" that cannot be dismissed as unreal simply because medical science cannot detect them. Furthermore, when science declares that my reports regarding such experiences are not objective, and are unreliable because they are subjective, this is merely an excuse to justify the dismissal of phenomena that medicine cannot easily concretize, categorize, or conceptualize under its present limited rules of acceptable evidence.

It is patently absurd to consider only matter and energy in medical decision-making while disqualifying universal human attributes such as thought, emotion, will, intent, dreams, desires, psychic phenomena, and consciousness itself. To cavalierly explain such phenomena away by calling them illusory or unreliable by-products of the physical brain amounts to a monumental unscientific liberty taken by those blinded by the biases of modern medicine. Alan Wallace points to the unacknowledged elephant in the room, the fundamental contradiction to the proposition that mind is an illusory phenomenon:

> *The central aim of science is to understand and control the objective, physical world; yet the subjective mind, with its powers of observation and reasoning, is, awkwardly, the fundamental instrument of scientific inquiry.*[11]

This type of objectivism that claims to be the arbiter of factual reality is deeply flawed and dependent on a circular form of logic. When up against the wall it defends itself by arguing that anything beyond physical existence is unworthy of scientific inquiry simply because it is subjective. It falls back upon its preconceived unscientific assumptions concerning reality in order to justify its scientific version of reality.

Clearly, when questioned individually and off the record, many scientists and physicians do not deny the reality of things like consciousness or psychic experiences. But within official channels, few dare to acknowledge such views let alone assert that they may have an impact on patient outcomes. And so, for all practical purposes, so-called subjective phenomena constitute inadmissible evidence in the court of material medicine.

Returning to the issue of pain, back pain in particular serves as a good example of the fickle nature of so-called objective evidence. If a person with back pain were to consult a physician, a physical exam may reveal some muscle tension and x-rays may show a bulging intervertebral disc. This, of course, is the solid evidence that medicine prizes. The problem is that there seems to be no real correlation between such objective markers and the experience of back pain. In other words, there are many who complain of back pain who show no evidence of reasons for pain either on exam or imaging tests. Likewise, there are many whose x-rays reveal evidence of bulging discs or arthritic deterioration of the spine, for example, even though those persons do not complain of back pain. Although the relationship between back pain and objective markers appears to be random, physicians, nevertheless, are inclined to point to those markers as explanations for pain when they are present.

When a person notes that his back pain began immediately upon waking from a dream of a tidal wave about to hit land, it tends to be dismissed as an interesting anecdote. When another person reveals that she has been under a great deal of pressure to lend assistance to an aging and infirm parent who has a history of being emotionally abusive, it is often written off under the general category of "stress." These types of subjective associations with pain are taken much less seriously than objective findings like degenerative discs and spinal narrowing. The risk for patients who show no objective indications of back pathology is that they may be told that their pain is not real. On the other hand, the danger of objective markers is that they may convince patients of the irreversible materiality of their pain.

Stress has become a convenient dumping ground for the so-called unscientific information that medicine fails to adequately address. Stress is acknowledged but not taken very seriously. In fact, stress is modern medicine's repository for all of that personal, subjective, anecdotal, qualitative, anomalous, intangible information that patients often yearn to reveal in the safe confines of a therapeutic relationship. Doctors often don't know what to do with such information. When in doubt, they may casually blame it on the vague generality of stress.

The implicit message is that such information is not central to the medical problem at hand. In the name of clinical purity, only that which can be detected by the methods and tools of medical science is considered pertinent. Sadly, although physicians are trained to demonstrate compassion, they sometimes give the impression of caring about stories that they secretly consider immaterial and beside the point. When patients sense such insincerity—one of those subjective human experiences—it can have a significant and adverse impact on the doctor-patient relationship.

The ostensibly pertinent facts, as determined by medical science's definition of reality, play an enormous role in the strategies employed in the misguided war against disease. Such facts are frequently at odds with patients' experiences, personal beliefs, and understandings of their own illnesses. One has to wonder about the effectiveness of a form of medicine that relies so heavily on objective information.

6. Objectivity as clinical detachment from patient experience

It is wise and beneficial to maintain healthy boundaries between doctor and patient. However, objectivity sometimes serves as an excuse for health care providers to maintain an unhealthy distance from the painful realities of their profession and of human suffering. It is important to be human and humane. Patients are less likely to respond with a sense of trust to a clinically detached technocrat than to someone who genuinely wishes to get to know them. Excessive clinical detachment often has the net effect of depersonalizing those whose illnesses would respond best to a physician's personal attention.

It would be near impossible to function in my own medical practice if I were not able to elicit patients' stories and their subjective impressions regarding their illnesses. I would not be able to accomplish this if, like Joe Friday, I was interested in "Just the facts, ma'am." I have come to realize that I can serve my patients best when I know them as one interested person to another, like a neighbor or good friend. Most patients are conditioned to expect busy clinicians who have just a few precious moments of time to listen to their presenting complaints. When my patients realize that I am genuine and that I wish to know more, their responses can range from pleasantly surprised to downright shocked.

Patients have assimilated the underlying messages of their physicians so well that they often come ready and willing, in the name of efficiency, to communicate in the medicalized language that they have absorbed through their interactions with the medical system. Some are so well programmed that they don't know what I mean when I ask them to describe their symptomatic experiences. They skip the step that involves empirical observation of personal symptom patterns and jump straight to preformulated diagnoses and proposed treatments.

One such patient might say, "I think I have a sinus infection, doc, because the mucus I blow out is yellow. Do you think I need an antibiotic?" rather than, "Over the past week I've developed a feeling of pressure above and below my eyes, and my nose has been stuffed up. Mucus runs down the back of my throat, and when I cough it up it looks yellow.

When I swallow, there is a sharp pain in my right ear. I've been under a lot of strain at work lately."

By ignoring patients' subjective observations of their illnesses, clinicians have taught patients to do the same. I frequently find myself having to educate patients how to observe the details and patterns of their symptoms. A person with asthma, for example, may report that he wheezes and has difficulty breathing. But when I ask what factors exacerbate or ameliorate the condition, he may be at a loss to respond. Only after given instructions does he then report at the next visit, for example, that his asthma flares up at 4pm, is often accompanied by indigestion, and tends to settle down by 8pm. Belching, he observes, brings temporary relief to his shortness of breath. Believe it or not, details of this nature are very important to the successful implementation of homeopathic treatment.

There is a big difference between making unprejudiced decisions and ignoring information because it does not conform to the narrowly defined parameters of orthodox medicine. The former is a legitimate goal, while the latter is a means of avoiding the true impact that patient experience has upon the etiology and course of disease, and the effects that medical interventions can have upon patients' lives. Medicine has a choice; it can continue to cling to the illusion of its clinical objectivity as the only true guide to managing illness or it can begin to open its eyes to the realities of individuals and their subjective experiences of suffering.

7. Objectivity as in value-free or value-neutral

Science famously claims to be independent from issues involving the making of value judgments. It is concerned strictly with the facts, or so it says, and professes to have little to do with issues of morality, of right and wrong. It is only concerned with ascertaining the factual truth. It does not take sides because it believes that those facts are not open to debate. Since science deals only in fact and does not make value judgments, it believes that its conscience is clean.

Science believes that it can achieve value neutrality by dissociating itself from all things human. It does this by declaring that the subjective dimension of human experience does not fall within the scientific domain. In yet another trick of circular logic, science argues that it is only concerned with objective fact, and is not influenced by human values, because values are subjective and not a part of science. Of course, this is a form of wishful thinking that many indoctrinated into scientific culture accept without questioning. The only way this could be true would be if science were to operate in a vacuum, as if its actions were not an extension of human activities, beliefs, desires, and foibles—as if humans were androids.

Medicine wishes to follow suit, claiming to be value-neutral, but this is a rather dubious assumption to make given the thorny subject matter of human health and illness. When pressed on the matter, most reasonable persons concede that absolute value-free objectivity is not possible in either science or medicine. Given this caveat, most still believe that science strives to be as objective as possible, and that it has been very successful at doing so.

Science achieves the illusion of value neutrality by compartmentalizing issues in before and after terms. Prior to and during scientific activity it assumes that values are not at issue. Scientists conduct investigations, which, by definition, entail the unbiased examination of objective phenomena. The results produced by science, however, are another matter. Once its activities are completed, science admits to a number of issues that can arise after the fact, most obviously having to do with the application of knowledge produced by science. In essence, science dissociates itself from the results of its activities in order to justify conducting those activities.

Sticky issues that emerge after the fact are conveniently delegated to ethicists to handle, thus absolving science of charges of being associated with subjective value judgments. Here is where science fails to take responsibility for its actions, claiming that the knowledge it produces is an issue for others to ponder over—as if to say that science cannot be judged for the creation of that knowledge or the consequences that it may bring.

The biotechnology industry is a prime example of how the scientific imperative, ostensibly pursued for the benefit of humankind, can remain undeterred from its researches, even after leaving a trail of destruction in its path. Having discovered genetic markers that can predict the likelihood of a variety of diseases, medical scientists are eager to make this new discovery available to the general public, regardless of the fact that most such genetic diseases are incurable. Never mind the psychological double bind that places patients in, knowing that they may contract a disease but also knowing that there is no effective treatment. It turns out to be good for business because a lot more people are going to need treatment for anxiety and depression. Having created yet another dilemma for humankind to wrestle with, medical science blithely moves on to the next "value-free" investigation, unscathed by the former.

Prior to scientific inquiry, there is little to no self-examination, no questioning of basic premises. Science somehow manages to successfully delegate ethical responsibility after the fact, while its basic assumptions are rarely ever questioned. On this count, science and medicine believe that they are in the clear, and cannot be accused of dealing in beliefs or issues involving value judgments. In my estimation, however, this is clearly not the case.

The foundations of both science and medicine are constituted of metaphysical propositions that involve distinct value judgments. They reflect the values of their paradigmatic worldviews, their own chosen values. Science and medicine choose to investigate the material, leaving the immaterial for non-scientists to contend with. Science prefers reductionist solutions, showing no interest in holistic principles. Medicine believes in phenomena that have cause and effect explanations, while conveniently ignoring the mind-body interactions that lie at the root of many illnesses. The choice to give preference to objective phenomena is, at bottom, a value judgment! When science discriminates against the subjective, it fails to meet its own standards of objectivity. Both science and conventional medicine take the same distinct metaphysical stance regarding the universe. That stance is defined by the same –*isms* that are used to justify the value neutrality of their intentions and actions.

Science can be value free only when it exists in a vacuum, only as a theoretical ideal—and even then it is debatable—but it is never value-neutral in the real world of actual people and their real sufferings. While bioethics may give the impression that medicine takes responsibility for its actions, medicine has yet to honestly examine its metaphysical presuppositions. Those presuppositions do not involve issues of scientific fact; they require choices that involve value judgments.

Finding Balance

Each of the preceding connotations of objectivity is biased in ways that have practical implications for the practice of medicine. Science is certainly not objective in the way that the average person understands it to be. Claims as to the objectivity of medical science are not supportable and, yet, medicine continues to promulgate this falsehood.

The irony is that when medical science boasts of its claim to objective factual knowledge regarding disease processes, it arrives at this conclusion via biased presuppositions. Medical science's objectivity is paradoxically a product of its own subjective assumptions about what constitutes valid and trustworthy scientific evidence. In the end, however noble medicine's intentions may be, its body of knowledge is critically flawed by virtue of the fact that it is deeply prejudiced in favor of what it calls objective data.

Objectivity, as it is currently defined by modern medicine, lends itself to dissociation from first-hand experience and isolation from the clinical realities of patients and their problems. Medical science is fast falling behind the paradigmatic changes that have been taking place in the world of physics for decades now. Failing to incorporate the lessons learned from the new quantum worldview, medicine continues to act as if our bodies can be treated as erector sets subject to the principles of Newtonian physics. More important, the new physics long ago exploded the myth of objectivity as an illusion based in false materialistic and mechanistic bias. And yet, the entire premise of medical excellence is built upon a scientific model that holds itself up as the epitome of objective methodology.

Medical research is highly regarded precisely because it is believed to be the embodiment of objective scientific method. I would contend, however, that this is a mischaracterization. Although genuine in its intent and conviction, most medical research is grounded in biased assumptions that blind its proponents from attaining a more balanced perspective. It is more accurate to describe medical research as abstract, quantitative, and rational in its approach. These qualities are mistakenly believed to be characteristic of objectivity.

Conclusions drawn by medical science do not automatically become objective simply because they were arrived at through rational thought and statistical analysis. Perhaps you are familiar with that maxim brought to us courtesy of the computer age: garbage in, garbage out. When fifty percent of the data represented by subjective experience is excluded from scientific investigation, we can only expect skewed results. The end result of so much unacknowledged bias is a sterile and unbalanced approach to health and healing that treats stereotypes of diseases rather than flesh and blood patients with hearts, minds, and souls.

Is absolute objectivity even possible? Most definitely not. True objectivity cannot be achieved except perhaps in matters such as mathematics. Everything else must be reduced to degrees of subjectivity. Of course, mathematics is a highly abstract symbolic language and is the furthest thing from organic life imaginable. Objectivity is a little more realistic in the context of the study of an inorganic science such as geology. However, the closer we come to the study of organic life forms, the harder it is to maintain an objective stance. The study of ourselves, of human life, health, and illness, is highly subjective in many respects and quite refractory to the dictates of objectivism.

Is objectivity even desirable? Some argue for a less stringent definition of objectivity, which allows for human error but is still focused on remaining as unbiased as possible. I have no problem with aspiring to objectivity in this very limited sense. Objectivity is a relative value, not an absolute ideal. The problem lies with our definition of biased. If one assumes that the inclusion of subjective factors constitutes bias, or that phenomena that cannot be quantified are not compatible with science, then I must disagree. There has to be a better way. Either the limitations of scientific investigation must be clearly defined and truthfully acknowledged for what they are, or the very nature of scientific methodology must be amended.

Western medicine considers objectivity to be a desirable ideal and genuinely believes that its methods have produced the most reliable medical knowledge to date. Information derived from objective sources is taken to be factual and irrefutable. I believe that it is a mistake borne out of dualistic thinking to say that subjectivity and objectivity are opposites, that they are negations of each other. The truth is that they are complements. While dualism conceptualizes them in opposition, holism understands them as two sides of the same coin. To eliminate one in favor of the other is to tell only half the story.

My own homeopathic medical practice is a perfect example of how objective and subjective information can have a very real impact on therapeutic choices and patient outcomes. Homeopathic methodology involves gathering both types of information, the totality of which is then subjected to a unique form of analysis. I have seen successful outcomes that were dependent upon objective information as in, for example, anemia reflected in blood work, or the confirmation of an ovarian cyst by ultrasound. Other successful prescriptions would not have been possible without knowledge of personal subjective details that could never have been detected by the tools of science, such as dreams of fire, fear of snakes, a craving for lemons, sensation of a lump in the throat, or anger triggered by the sound of someone chewing their food. All represent clues that can point directly to specific homeopathic solutions.

Medicine's allegiance to objectivity has resulted in the unwarranted elimination of a great deal of information that could otherwise have an enormous impact on human health. When we objectify something we leave out the personal, and when we strive to be scientifically objective we leave out the subjective. The recounting of one's personal story of suffering to an attentive and sympathetic listener may very well be the most important step in the initiation of the healing process. Since medicine has predetermined that this type of information is of little use, it places itself at an unnecessary disadvantage right from the start.

A more balanced approach does not shy away from subjective criteria simply because they are difficult to quantify. A balanced scientific methodology accounts for both objective and subjective phenomena. Although they do represent different types of information, one is no more important than the other. Together they form a more complete picture, a more accurate rendering of holistic reality.

CHAPTER 7

Medical Empiricism

We have to conclude that science did not start from experience; it started by arguing against experience and it survived by regarding experience as a chimera.
    –Paul Feyerabend, *The Tyranny of Science*

Each moment of experience is the intersection of the past, present, and future, the convergence of the inner and outer.
    –Don Salmon & Jan Maslow, *Yoga Psychology*

I must admit that I had originally struggled quite a bit with this particularly perplexing *–ism*. The conventional definition of *empiricism* never made sense to me. But then one day it dawned on me that the problem stemmed from my own unusual perspective. Having practiced an empirical form of holistic medicine for many years, I naturally assumed empiricism to be something that it is not, at least not in the eyes of most mainstream scientists and physicians. Empiricism, it turns out, is a term that can take on different meanings, depending upon how you look at it.

Empiricism in general is a broad topic with little agreement among its proponents, except perhaps when it comes to its modern scientific proponents—they appear to believe that they are in accord regarding its definition. But even among contemporary thinkers, there seems to be a number of significant contradictions as to what is meant by the word *empiricism*. To wade into a comprehensive discussion of empiricism and its variants, both historically and scientifically, would be to open a giant can of philosophical worms. Upon closer scrutiny, *empiricism* and *empirical* turn out to be ambiguous terms that tend to suit the needs of those who employ them. Suffice it to say, there are many inconsistencies of meaning and a great deal of confusion regarding the topic. For my purposes here, I shall confine the discussion mainly to medical empiricism and its relationship to a more holistic understanding of empiricism.

One popular definition of science goes something like this: *science is a systematic method of testing hypotheses against empirical evidence obtained through observation.* A shorter version would be: *knowledge gained through observation and experiment.*

Very few if any scientists would dispute these characterizations. They believe science to be an empirical method of observation and experiment. Empiricism, by the way, is the only *–ism* explicitly and consistently used in definitions of science. The confusion, however, arises from what is meant by empiricism. It is my contention that modern medical science no longer adheres to its own definition of empiricism. Neither is it aware of this internal inconsistency regarding one of the important foundational principles of scientific discovery. It will take a little time to explain what I mean.

Empiricism is an epistemological theory of knowledge. Epistemology is the branch of philosophy that deals with how we know what we know and how to distinguish knowledge from mere opinion. Historically speaking, empiricism was a philosophy debated among many intellectuals, most notably philosophers of the 18th century Enlightenment. John Locke, George Berkeley, and David Hume were the primary exponents of empiricism, a theory of knowledge that placed greater emphasis on knowledge gained from experience as opposed to the ideas produced through reason. As such, empiricism was often contrasted to, and believed to be in conflict with, rationalism.

The rationalism of Rene Descartes, on the other hand, set reason above and apart from all other faculties. The thinking power of the mind was assumed to be superior to feeling, sensing, intuiting, and spiritual experience. Descartes believed that the rational function was superior because the information it produced was thought to be more reliable and certain. Other forms of experience were not to be trusted. Even the senses of touch, smell, hearing, taste, and sight were considered unreliable and potentially deceptive.

This turns out to be more than a bit ironic since science prides itself on its empirical methods of observation and evidence gathering before formulating and testing its hypotheses. Empirical evidence is supposed to be unambiguous because it is directly perceived by the senses. Nevertheless, the Cartesian thesis has gone unchallenged within scientific circles for centuries. It has become ingrained in the scientific mind to such a degree that, when compared to analytical reasoning, most other forms of human experience are considered virtually irrelevant to the scientific process.

It was the contention of empiricist philosophers that all knowledge is derived from experience, from things that we can touch and hear and see. Empirical knowledge is knowledge based upon observation and experience rather than theory or logic. In fact, the word *empiricism* derives from the Greek *empeiria*, which translates into Latin as *experientia*. This, of course, is where the English *experience* comes from.

A fascinating piece of history reveals that there was an ancient faction of Greek medical practitioners who called themselves *Empirics*. Physicians of the Empiric school based their treatments on practical experience rather than the rational theories and doctrines of the Dogmatic school of the time. Also known as the *Hippocratici*, the *Dogmatics* were followers of Hippocrates and subscribed to the theory of the four humors, a concept first introduced into medicine by Hippocrates. This theory of disease is notable for having supplanted a prior belief in the supernatural causes of illness. It put a wedge between medicine and religion and introduced philosophy into medical thinking. Its followers were so enamored of the new rational approach that they attempted to discredit those who were not in agreement. The Empirics, therefore, were labeled quacks because they favored what their experience taught them over the new rational theories. For centuries, most physicians counted themselves as members of one or the other of these ancient medical sects.

Dogmatics believed it was necessary to understand the underlying causes of diseases before they could know how to treat those diseases. They searched for *hidden* causes by dissecting the body in order to understand its internal parts and how disease affected those parts. Here, we see the reductionist influence in medicine. Empirics, on the other hand, believed that, in the final analysis, all causes are proximate causes. Final, ultimate, or hidden causes are unknowable because it was thought that Nature as a whole will always remain unfathomable. Empirics believed in the need to understand *evident* causes of disease, that which was evident as a result of direct practical experience in treating sick individuals. They focused on their observations of nature, human sickness, and the outcomes of their treatments. This type of methodology is more closely allied with a modern holistic approach to healing.

Empiric physicians emphasized the practical outcomes of their interventions while Dogmatics concerned themselves with what they believed to be the first causes of illness—causes that were theorized to originate from the interior of the physical body and from individual parts of the body. Empirics valued empirical outcomes above all else, while the other school used logic and reasoning to draw conclusions regarding the origins and treatment of disease.

Amazingly, this philosophical rift persisted for hundreds of years and continues even to this day. It comes down to a question of whether physicians should rely more on the lessons learned from first-hand experience or abstract rational theories designed to explain disease and its treatment. This deep conflict within medicine was expressed two hundred years ago most famously by the founder of homeopathic medicine, Dr. Samuel Hahnemann, who did not hide his strong preference for medical empiricism when he wrote the following:

> *The physician's calling is not to make countless attempts at explanation regarding disease appearances and their proximate cause holding forth in unintelligible words or abstract and pompous ex-*

pressions in order to appear very learned and astonish the ignorant, while a sick world sighs in vain for help. Of such learned fanaticism we have had quite enough. It is high time for all those who call themselves physicians, once and for all, to stop deceiving suffering humanity with idle talk, and begin now to act, that is to really help and to cure.[12]

Hahnemann clearly sided with the lessons learned from experience and distrusted theories that were not validated by practical outcomes. So what gives? If the consensus view holds that present-day scientific method is a combination of rational theory, empirical observation, and experimental verification, then why were the empiric medical practitioners of ancient times disparaged as charlatans? Not surprisingly, the answer lies in how we define empiricism.

Let us review. A key feature of science is that it depends upon empirical observation. Empiricism is a philosophy of knowledge that emphasizes knowledge gained through experience. But the experiences of Empiric physicians did not qualify as legitimate and somehow made them unscientific quacks, at least according to the opposing medical camp of the day. Dictionary definitions of *empiric* confirm this truly curious and contradictory assertion. Here are entries from several different sources. The first is from a standard dictionary:

> *Empiric: A person who, in medicine or other branches of science, relies solely on observation and experiment. A quack doctor.*[13]

The second is from an older medical dictionary published in 1917:

> *Empiric: Based on practical observation and not on scientific reasoning. One who in practicing medicine relies solely on experience and not on scientific reasoning. A quack or charlatan.*[14]

Here are definitions from two different modern medical dictionaries:

> *Empiric: A practitioner whose skill or art is based on what has been learned through experience.*

Empirical: Based on experience rather than on scientific principles.

Empiricism: Experience, not theory, as the basis of medical science.[15]

Empiric: A member of a school of Graeco-Roman physicians, late B.C. to early A.D., who placed their confidence in and their practice purely on experience, avoiding all speculation, theory, or abstract reasoning; they were little concerned with causes or with correlating symptoms in order to gain a true understanding of a disease; they even held basic knowledge, physiology, pathology, and anatomy in low esteem and of no value in practice.[16]

The crux of the matter, it turns out, is that the definition of empiricism depends upon how we define experience. Empiricism as it is defined by medicine, then and now, is an empiricism of the physical senses—that which can be heard, smelt, felt, tasted, and touched. The experiences of Empiric physicians did not qualify because the Dogmatic or rational school of medicine limited experience to sensory experience. It did not include the contents of mind. It excluded that which most people assume to be the bulk of human experience—consciousness. The conventional notion of empiricism accounts only for sensory experience and leaves no place for mental experience.

The irony, of course, is that there would be no science at all without conscious human participation, without the activities of the mind. It seems as if we have exposed a particularly bold deception in the way that scientific method is portrayed. It claims to be empirical but rejects that other version of medical empiricism by defining empiricism in its own materialist terms. The only experience that counts, therefore, is objective sensory experience of the physical objects that constitute the material universe. Here, we find a remarkable inconsistency in that the historical antecedents of holistic practitioners were derogatorily called Empirics, while medicine simultaneously claimed then, and continues to claim now, that its methodology is empirical.

There is yet another problem inherent in this awkward and contradictory characterization of medical empiricism. If conventional medical empiricism is based on observations of physical objects made by the human senses, then how do we account for the notion that sensory experience is, first and foremost, subjective in nature? After all, people often perceive the same object quite differently. The taste of a food item, the sound of an orchestra, the feel of a garment, and the visual perception of a film are subject to the interpretations of the individuals experiencing them. Such experiences can be described in a colorful variety of ways, so much so that people sometimes wonder whether they are describing the same event. Need we be reminded that the most ubiquitous and important dimension of human illness, pain itself, is an almost purely subjective phenomenon?

How is it possible, then, that medicine would consider sensory experience legitimate while dismissing mental experience as unreliably subjective? Is it not a contradiction to say that sensory experience is misleading while at the same time defining science as a fundamentally empirical methodology? The answer is that medicine does not make this claim. In another twist of rational license, medical science concurs that there *is* a subjective aspect to sensory experience, which it finds to be untrustworthy. However, it also claims that there is a more reliable, objective component to sensory experience.

It should not be surprising that information obtained via the traditional five senses is considered to be mostly subjective. But there is an inherent and significant contradiction in the notion that scientific data must first be observed by the subjective senses before it can ultimately be transformed into objective information via the instruments of science and calculations of the rational mind. The information obtained, for example, from viewing a tissue sample with the subjective naked eye, somehow becomes scientifically acceptable objective information when viewed by the same eye through the lens of a microscope. The irony here is inescapable.

Like all other phenomena interpreted from a left-brain perspective, sensory experience, it is claimed, has two separate components, the subjective and the objective. In accordance with this dualistic perspective, medicine splits our experience of the world into two—the objective, which is real, tangible, reliable, and measurable, and the subjective, which is illusory, immaterial, unreliable, and unquantifiable. In making this claim, medical science restricts the definition of empiricism to include only the so-called objective dimension of sensory experience.

This obviously begs the question, what is objective sensation? The objective dimension of sensory experience, it turns out, translates into the anatomical and quantifiable components associated with our subjective sensory experiences. Visual experience, therefore, becomes concrete, objective, and scientifically permissible when it is spoken of in terms of structures of the eye and areas of the brain that interpret neural impulses coming from the eye. Color, shape, texture, depth, and movement become rods, cones, pupils, corneas, optic nerves, visual cortexes, and electroencephalogram readings. Likewise, the perception of subjective auditory phenomena like pitch, loudness, and timbre are reconstituted as tympanic membranes, ossicles, auditory nerves, and sound wave amplitudes and frequencies.

This brings to mind another sticky issue. It is impossible to accurately characterize sensory experience, even in anatomical terms, without that subjective phenomenon called human consciousness. Is there such a thing as objective sensation or is this just a distinction made for the convenience of the medical mind in order to eliminate the cognitive dissonance that accompanies such a notion? A person's experience of a particular singer's soft, feminine, velvety voice is going to be vastly different from a scientist's description of that same voice. Is it not a contradiction to deem a certain aspect of sensory experience untrustworthy when all experience is, at bottom, subjective? If scientists can convince themselves that there is an independent objective dimension to sensory phenomena, then they can avoid violating one of the main tenets of scientific methodology, which is that subjective information is unreliable and of little value to scientific inquiry.

The universal dimension of personal experience is excluded because it does not qualify as empirical evidence, thereby rendering it virtually irrelevant when it comes to scientific deliberation. Medicine frequently rejects personal experience and redefines it to conform to the various –*isms*—objectivism, materialism, reductionism, and so on—in order to make it more scientifically acceptable. As such, it becomes a self-fulfilling methodology that reinforces its presuppositions regarding the nature of health and disease.

Let us return now to our definitions of an Empiric physician. It is stated that an Empiric is one who relies *"solely on experience," "solely on observation and experiment," "rather than on scientific principles,"* and *"not on scientific reasoning," "not theory,"* thereby *"avoiding all speculation, theory, or abstract reasoning."* This composite definition should strike one as rather incongruous given our understanding of scientific method. It seems to be saying that science is not science without speculation and abstract theories, and that a method that involves experience, observation, and experiment cannot be scientific without an accompanying speculative theory to back it up. One can get twisted up in knots trying to make sense of such contradictory double-talk.

It is my belief that this illogical claim to scientific logic is nothing more than ancient medical politics carried down through the decades. This medical war between Empirics, who are the analogs of modern day holistic practitioners, and Dogmatics, who are the predecessors of contemporary proponents of mainstream rational medicine, has been going on for generations. One camp relies primarily on empirical observations made from the trial and error treatments of its patients, while the other occupies itself with theoretical causes and rational explanations of disease.

The rational school, which puts most of the emphasis on logic and theory, has maintained dominance over time not because of the superiority of its methodology, but primarily through persuasion, politics, and power. There is something deceptively appealing about a person's ability to engage in logical argument, polemics, and sophistry. It impresses a lot of people even though it says little about the knowledge, experience, wisdom, or rightness of a person's position. Nevertheless, such skills contribute to the indisputable popularity of rationalism.

The knock against Empiric physicians was that they relied *solely* on their experiences and did not use reason while treating patients. Of course, it is a patently absurd notion that such physicians did not use their rational faculties to draw conclusions from the outcomes of their treatments and then use that information to amend their practices over time. After all, that is the very nature of trial and error. This also did not preclude them from formulating ideas regarding the nature of illness and its cure. What they did not condone was the notion that theory should trump concrete results. To adhere to a theory of disease in the face of evidence to the contrary would amount to ideological foolishness—like repeatedly subjecting a sick patient to bloodletting while his or her health status slowly declines over time.

It is the same rationale that justifies contemporary medicine's reluctance to entertain a variety of holistic therapeutic alternatives. When alternative systems of healing do not see eye to eye with conventional medical principles, it is believed that they must be in error or based on faulty premises. They are sometimes accused of defying the laws of biochemistry, physiology, pathology, and so on. Holistic practitioners respond to skeptics by asking that they suspend their judgment and preconceptions regarding disease in order to examine the results and listen to reports of patients who have undergone holistic treatment. Diehard skeptics have been known to dig in their ideological heels, employing a variety of tactics to deny the evidence, including the rather unconvincing argument that any success achieved should be dismissed as a manifestation of the placebo effect.

It is untrue that most empirically-oriented therapies do not have theory, knowledge of disease, or reasoning to back them up. The historical reality is that those unconventional systems of medicine and healing did not agree with the medical theories of the day. Empirics were vilified because they disagreed with the theories and methods of the medical establishment—and the same dysfunctional dynamic between alternative and mainstream medicine holds true to this day. The historical lesson is that medical science is not considered science unless it conforms to the prevailing theories of the time.

In my mind, it should be the other way around. It is foolish to adhere to a theory in the face of experience that tells us otherwise. Furthermore, it is an illusion to believe that one camp should win out, should prevail over the other, as if there is no place for cooperation. It is far more reasonable to conclude that some medical problems are amenable to one approach, other situations may best respond to the other, and many require both. Absolute reliance upon one philosophy or the other—empiricism or rationalism—is an exaggeration perpetuated by politically polarized camps.

Historian, Harris Coulter, has documented in great detail the long-standing medical disagreement between empiricist and rationalist factions. In fact, he notes, it is a theme that has recurred since the dawn of Western medicine. In his book, *Divided Legacy: A History of the Schism in Medical Thought*, Coulter contrasts empiricism's emphasis on the practical treatment of patients with rationalism's preoccupation with theory and logical appeal to physicians:

> *The difference between the two doctrines can be discussed in terms of the relations between theory and practice. ...The idea that therapeutics is primary, and itself yields knowledge of the organism, is one of the great Empirical contributions to medical philosophy.*[17] *...Empiricism is a set of rules for medical practice. It is a method of treating the sick organism and thus of investigating it. Its function is purely curative, and this ideology is oriented toward the patient—his need to be cured as skillfully, rapidly, and carefully as possible.*
>
> *Rationalism is a theory of the organism. Its function is to explain the workings of the organism in sickness and health, and it is oriented toward the needs of the physician.*[18] *... Professional cohesion is engendered by doctrinal coherence. Hence the desire for logical consistency is a hallmark of Rationalism. These physicians stressed theory over practice because a convincing logical theory seemed more important to them than success in practice.*[19]

Here, Coulter astutely recognizes that these divisions are artificial, exaggerated, and not as cut-and-dried as they appear to be. Furthermore, the two camps are a natural consequence of psychological types or, as I have framed it, the different modes of perception of right and left-brain dominant individuals.

> ...medical history represents the interaction between the Empirical and Rationalist systems of thought. This polarity is perennial, recurring in each new age clothed in the medical vocabulary of that age.
>
> All physicians do not adhere strictly to one or the other system. The Empirical end of the spectrum will always contain individuals who veer toward Rationalism, while the Rationalist end will contain some who are moving toward Empiricism...
>
> The perpetuation of either system, and its waxing and waning, are determined by the recruitment to the medical profession of persons of the Empirical or Rationalist psychological type. Representatives of both types are found in every population, and different sets of social, economic, legal, and other conditions will encourage one type or the other to undertake the practice of medicine.
>
> Thus the appeal of the two systems is ultimately psychological—one reason (inter alia) why therapeutic experience rarely seems to convince the adherents of either system to abandon it and embrace the opposing one.[20]

We happen to live in an age that heavily favors left-brain rationalist thinking. The Scientific Revolution ushered in many remarkable achievements, but it also marked the onset of a deep and abiding mistrust of right-brain perspectives. While true holism is by its very nature inclusive, and therefore values both left and right-brain modes of perception, scientific rationalism is a left-brain perspective that sees the world in black and white terms and has little tolerance for opposing viewpoints—thus the antagonism toward empirically-oriented holistic modalities over the past several hundred years.

Homeopathic medicine is heavily influenced by empiricism. Its paradoxical methodology was formulated only after having accidentally observed its principles in action: quinine bark, known to be able to cause fever and chills, was successfully used by practitioners of the conventional school to treat patients with the fever and chills of malaria. This phenomenon was noted and then emulated.

Homeopathic researchers administered small doses of substances to test subjects in order to study their effects. Patients whose symptom patterns matched the symptom pattern capable of being produced by a given substance were then given small therapeutic doses of that matching substance. The prescription was then adjusted depending upon the results. Theories regarding health and illness were only formulated later, as a consequence of the empirical results observed. Thus, observations were made first, trial and error methods were developed next, and theoretical conclusions were drawn afterwards.

By contrast, modern medicine tends to first construct theories of causation, which subsequently determine courses of therapeutic action. Peptic ulcers, for example, are assumed to be either a function of excess stomach acid or an overgrowth of *Helicobacter pylori*, and those premises guide treatment. Likewise, depression is thought to be caused by various biochemical mechanisms that result in a deficiency of serotonin, a brain neurotransmitter. Treatment, therefore, involves prescribing antidepressants that are designed to manipulate serotonin levels in the brain. These are examples of how theory precedes practice. It is the hallmark of rational medicine. In this sense, it is the opposite of empiricism.

The heart of the matter ultimately rests in empiricism's exclusion of subjective criteria. In one of those rationalizing twists of logic, science decided early on that, even though empiricism is based in experience, experience as defined by the average person on the street is not true empiricism. Experience was redefined and reduced to objective experience, a conceptual oxymoron fabricated by the scientific mind to ensure the purity of its methodology.

Conventional empiricism only allows for the anatomical equivalent of reason, imagination, intuition, emotion, and so on. It addresses the supposed material location of consciousness—a slab of gray and white matter contained within the skull—while denying the reality of mind. But the brain does not experience phenomena; people do. Medical science spends a great deal of time studying the physical brain, but gives little thought to consciousness, to human life itself. Empiricism's concept of experience excludes consciousness, a phenomenon that has never been localized anywhere in the brain. And yet, consciousness appears to precede all, including that which we call objective knowledge. If consciousness contains only the subjective, then all of science would have to be reclassified as subjective because there is no science without the minds of the scientists who conduct it.

Of course, one may wonder how the study of consciousness is pertinent to our topic—health, illness, and healing. When medical science excludes the subjective mind from nature, it severs the mind's relationship to physical phenomena. The connection between mind and body, between consciousness and matter, is taken to be a given in most alternative therapies—it is a fundamental principle of holistic theory and practice. To deny the import or existence of psychic phenomena is to deny the power of the mind and its potential role in healing.

The reverse is also true—physical illness often leaves its mark on mental, emotional, and spiritual wellbeing. It is even more difficult to accept the notion that physical and mental symptoms often emerge together simultaneously, that is, as part of a pattern without a cause and effect relationship. Nevertheless, these not so coincidental synchronistic connections can be significant sources of meaning to patient and practitioner, and may inspire practical therapeutic interventions. Conventional medical empiricism precludes the scientific study of these important relationships.

In addition to the extant body of information produced by centuries of contemplative tradition, there is a formidable philosophical precedent that supports the notion of a scientific study of consciousness. William James was a nineteenth century physician and psychologist, in addition to being one of the most important philosophers in American history. He studied all aspects of human consciousness including religious experience. Alan Wallace has called James "a modern pioneer of the scientific study of the mind."[21]

William James developed a perspective that he called *radical empiricism*, which proposed the inclusion of all mental phenomena within a scientific framework of investigation. Rejecting dualism, he did not see body and mind as two separate and distinct phenomena. He proposed that the world did not consist, as Westerners are inclined to believe, of matter and consciousness, but rather of pure experience. He believed that all phenomena, material and immaterial, could be explained as a function of experience. It was the mission of radical empiricism to explore and examine this new frontier of pure experience. Wallace explains:

> *Eventually defining himself as a pragmatic empiricist, he came to the conclusion that science should always concern itself with direct experience, including the first-person experience of the mind. In other words, personal experience regarding mental phenomena was crucial. This assertion signaled a major expansion of the subject matter usually studied by empiricists. ...James also believed that introspection, the systematic exploration of one's own inner experience, could become the foremost tool for understanding the mind. ...James' belief amounted to reestablishing the presence of mental phenomena in the natural world. For him, the natural world—and recall that early on scientists were called "natural philosophers"—should include both the objective, physical world and the subjective, inner realm of the mind.*[22]

James' more inclusive view of empiricism is highly compatible with holistic medical theory. Any perspective that fails to move us beyond the strict material limits of conventional health care is flawed because it is incapable of answering some of the more important questions, especially questions as to the roles of purpose and meaning in the generation of human suffering.

Now let us recap with the intent of highlighting some of the contradictory logic behind medical empiricism:

1. Medicine rejects empirically-based therapies while at the same time claiming to be empirical.

   - Science defines itself as an empirical method of observation and experiment.
   - Empiricism is a philosophy that prioritizes the knowledge gained from experience.
   - Empiric physicians relied primarily on their practical experience to treat patients.
   - But the experiences of Empiric physicians did not qualify as empirically valid in the eyes of Dogmatic physicians.
   - Empiric physicians were criticized because they were guided by practical experience and not by theoretical concepts and the hidden causes of disease.

2. Medicine downplays the common everyday understanding of sensory experience in order to objectify experience.

   - Science prides itself on objective empirical evidence that is supposed to be unambiguous because it is directly perceived by the senses.
   - But information obtained via the traditional five senses is considered to be mostly subjective.
   - To mitigate this inconsistency medicine focuses on the objective anatomical correlates of subjective sensory experience.
   - Science does this even though it is impossible to identify the specific nature of subjective sensory impressions by studying the anatomical correlates of sensation.

3. Medical empiricism acknowledges only the objective brain even though it requires the subjective mind to formulate medical theory.

- Conventional empiricism leaves no place for subjective mental experience.
- The only acceptable experience is objective sensory experience of objects of the material universe.
- While the physical brain is of great significance to medicine, consciousness is at best an afterthought.
- And yet it is impossible to formulate theories of disease or treatment without the aid of the subjective mind.

Confused? Well so was I, until I realized one day that it makes no sense other than as an elaborate network of rationalizations used to justify one of the medical –isms. The scientific definition of empiricism quickly falls apart under careful scrutiny. It is a product of several tricks of logic designed to give an appearance of congruency to the internal inconsistencies inherent in modern medical science's conception of itself. Empiricism has evolved into the twisted knot of logic that it is in order to keep pace with contemporary scientific medicine.

The rationale for conventional empiricism is an example of what happens to an idea when it is manipulated to conform to the ideological principles of Western medical thinking. By excluding all mental phenomena it adheres to materialistic principles. In order to remain objective it must exclude the subjective dimension of sensation. When defined in terms of the anatomical parts associated with sensation, it remains consistent with reductionism. And the overall logic of empiricism itself is a testament to rationalism at its abstract best.

It is a disturbing reality that modern medicine believes it can heal illness without consideration for the role of consciousness. Medicine disregards the powers inherent in thought, emotion, dreams, imagination, intuition, will, meaning, and belief in order to remain faithful to its distorted ideal of scientific purity. Empiricism, if it is to be redeemed, must be expanded to include both the subjective and the objective. We must return to a more humane understanding of experience that includes the contents of mind.

CHAPTER 8

Medical Conformism

The plague of mankind is the fear and rejection of diversity: monotheism, monarchy, monogamy and, in our age, monomedicine. The belief that there is only one right way to live, only one right way to regulate religious, political, sexual, medical affairs is the root cause of the greatest threat to man: members of his own species, bent on ensuring his salvation, security, and sanity.

–Thomas Szasz, *The Untamed Tongue*

T his particular *–ism* is rather obvious in some ways but not so apparent in others. One of the lesser critiqued aspects of Western medicine is its insistence that scientific method be used to discover *generalities* that can be applied across the board to as many people as possible. Contrary to what medical science leads us to believe, this is not an intrinsic feature of scientific method. It is orthodoxy's co-optation of scientific method for the sake of convenience to meet its own need for broad applicability, homogeneity, and conformity. Medicine views unorthodox diagnostic and therapeutic modalities unfavorably because it is believed that they cannot be uniformly administered to large groups or broadly conceived categories of patients.

The overarching trend in conventional medicine is toward generalization. It begins with specific instances and, from there, tries to draw broad conclusions regarding larger populations. The same idea applies in reverse; general principles are formulated and then assumed to be compatible with specific instances. Medicine does this because it values predictability and control. In a similar vein, the one most important criterion that gives credibility to medical research studies is that their results should be reproducible. It is assumed that a diagnostic tool, therapeutic intervention, or medical principle is worthy only if it applies consistently to large groups of individuals. As a consequence, medicine ignores anomalies in order to focus its attention on that which it perceives to be the norm.

In an inherently messy and unpredictable world, it is only natural to wish for order. It is human nature to want to reduce the complexities of life to more manageable terms. Medical practice is made a whole lot easier by grouping illnesses into neatly defined categories. It is all the more desirable if a single treatment can be applied to a large number of people in a given disease category. But these aspirations inevitably come into conflict with one very big issue—holistic reality. Human health and illness is highly complex and not amenable to oversimplification. Although this presents problems for cookie-cutter medicine, it is the same factor that makes holistic healing challenging, fascinating, and never boring.

Undaunted, medicine pushes on relentlessly in search of uniformity. Scientific method rightly looks for commonalities among many diverse phenomena, and this can prove to be useful when it comes to working with human suffering. But it can also be a hindrance. Medicine wrongly seeks to impose this imperative upon most, if not all, instances of illness and wrongly assumes that it is desirable to do so. In this sense, the method takes precedence over the goal of healing the sick by all means possible.

It is helpful for medical science to organize and group symptoms, diseases, lab findings, and treatments in ways that make them easier to understand and manage, but much valuable information is lost when nuance and complexity are sacrificed for the sake of simplicity. In the end, it represents a form of dumbing down, a paint by number approach that provides a false sense of certainty and predictability.

A diagnosis of eczema usually calls for the routine prescription of an industry standard topical steroid. It requires more effort, however, to recognize that identifying eczema is only the first step in what should be a more comprehensive evaluation process. It is critical to understand each case of eczema within the context of the individual and all other spiritual, mental, emotional, and physical health issues that he or she may be grappling with. Only then does a truly appropriate and effective short and long-term plan of action become possible.

Some cases of eczema are triggered by environmental allergens; some are not. Some cases flare up with emotions like anger, anxiety, or grief. Sometimes eczema can alternate with other medical conditions. For example, as eczema recedes, asthma or arthritis may flare up, and vice versa. In such cases, treatment of the eczema by suppressive means is not advisable. Suffice it to say that each individual case must be handled in accordance with its unique circumstances.

But the need for uniformity usually trumps the needs of patients and, so, topical steroids tend to be prescribed regardless of the context. Although diagnostic categorization and therapeutic standardization serve to satisfy the overwhelming demands for medical conformity, they do not necessarily yield the desired patient outcomes that we would wish for. The long-term health and satisfaction of patients has been usurped by the methodological need for simplicity, convenience, and uniformity. Concerns for methodological purity wind up overshadowing issues of treatment effectiveness.

The medical needs of individual patients should not be judged solely by the diagnostic categories imposed upon them. Conventional diagnoses are useful but not the be all and end all of the medical evaluation process. Likewise, the value of a particular therapy should not be based on its broad applicability. It should be considered no more or less valuable than some less commonly indicated approach that may be of benefit to a smaller number of cases.

The objective of medical science should not be to discover one-size-fits-all drugs while ignoring other, more individualized, non-pharmaceutical approaches to similar problems. It is pure bias and mere convention to believe that the only useful therapies must come from corporate pharmaceutical industry behemoths rather than from, for example, the indigenous practices of a thousand year old healing tradition.

Medicine prides itself on having standardized the entire enterprise of medical education and practice. The various specialties and subspecialties are organized with bright lines of demarcation, so much so that practitioners are reluctant to step over those bounds for fear of treading on forbidden turf. Specialty boards certify qualifying candidates, who are then entitled to be experts who have the final say on medical issues that fall within the scope of their expertise.

These relationships are set in virtual stone and may carry legal implications for those who have something of value to offer but who cross those boundaries in the process. Courts of law take it for granted that the specialist in neurology, for example, is the highest authority when it comes to a condition like multiple sclerosis. This precludes the legitimate recognition of a variety of unconventional practitioners who may have a great deal of knowledge not normally part of a neurologist's training that can be of practical value in treating multiple sclerosis. This quasi-legal hierarchical arrangement serves the conventional medical profession's interests more than it does patients.

Medical culture historically fashioned itself after early models of industrial productivity and military regimentation. In recent decades, medicine has sought to emulate contemporary standards of corporate culture. The common thread that connects these cultures is their homogeneity and their eradication of anything that would stand out as different or unconventional. Individuals sacrifice their personal preferences in exchange for business attire, military uniforms, and white coats and green scrubs. They follow strict codes of conventional behavior that are enforced by the sheer weight of peer pressure. On a collective level, quarterly reports and profit margins matter more than individual concerns. Workers assimilate the message that overall growth and productivity stand above the welfare of the individuals that make up the corporate enterprise. While such codes of behavior may accrue to the benefit of an army or corporation, they can easily stifle free and innovative thinking in a medical setting.

Cultures of this nature are loath to change their ways and do not readily incorporate new ideas. It should be easy to see why such an arrangement is not conducive to the well-being of suffering individuals whose illnesses do not conform to one-size-fits-all diagnostic categories. It is my belief that a great deal of dissatisfaction with the conventional medical model arises from forcing patients' complaints to conform to artificial diagnostic categories and therapeutic regimens that are not befitting of the true nature of their maladies. In short, they are victims of a homogenized medical culture that is too rigid to adapt to individual idiosyncrasies.

Corporate medicine assumes that it is desirable to merge our illnesses into diagnostic categories that necessitate predetermined pharmaceutical and surgical solutions. Homogeneity is the preferred norm. Approaches that take into account the individuality of persons and their illnesses are frowned upon because they cannot be efficiently incorporated into the streamlined objectives of the medical-industrial complex.

The imperative to generalize, however, flies in the face of clinical reality, which teaches us that there are as many permutations to human illness as there are people. It follows, of course, that truly effective treatments must be diverse and flexible enough to account for these variations. In fairness, the same critique can be leveled against many natural therapeutic interventions, which are often promoted and marketed as cure-alls for anyone willing to give them a try.

The whole modern concept of medical diagnosis is a manifestation of the drive for homogeneity. Like the taxonomic classification of zoological life forms, doctors are trained to believe that illnesses should match up with stereotyped disease descriptions and categories. Medical diagnosis is predicated upon a formulaic approach that results in the choice of a diagnostic code number taken from a book called the *International Statistical Classification of Diseases and Related Health Problems* (ICD) published by the World Health Organization (WHO). If it is not in the book then it does not exist, at least as far as insurance companies are concerned. The problems of real patients, however, rarely fit these neatly arranged categories of disease taxonomy.

More often than not, for the sake of insurance coverage for my patients, I choose to play the game by squeezing their presenting complaints into diagnostic codes that poorly approximate the true nature of their ills. The codes that I choose are often oversimplified caricatures of the actual real-life patients that I see in my office. This system of diagnostic coding is really just a game that doctors play to ensure insurance coverage for their patients, and insurance companies play in order to deny coverage.

The larger problem, though, is that the artificial classification of disease has an insidious effect on the practice of medicine. Physicians are seduced into believing that people's ailments are accurately represented by these categories and consequently look to fit their patients into them, even when they do not really fit. Clinical experience has taught me that, when the whole person is taken into account, very few ailments resemble the stereotyped versions described in medical textbooks. That is, unless I am willing to leave out a lot of information and focus only on a particular grouping of symptoms and some lab results that fulfill the diagnostic criteria of a particular disease category. In other words, I can only apply a diagnostic code with confidence and certainty if I subscribe to the reductionistic, materialistic, and mechanistic biases of the conventional medical worldview.

I have examined patients whose asthma symptoms are clearly allergic in nature, acting up only during springtime blooms. Some cases of allergic asthma only flare in the fall, and some are active continuously from spring through fall. Many are aggravated by dog hair, others by cat hair, and some by musty basements. There are others whose asthma only acts up when they catch a cold or an upper respiratory infection. Some asthma cases involve wheezing, some only manifest as coughing, and some involve throat constriction. Many cases are strictly exercise-induced. Other cases are highly psychosomatic—they are triggered by emotional states and psychological stressors. Some cases are accompanied by anxiety, and others by depression. Many cases occur only at night on waking from sleep. Even then, some occur predictably at midnight, others at 2 a.m., and still others at 5 a.m. There are cases of asthma that are triggered by damp conditions, others by cold dry winds, and some by hot humid weather. There are even cases of asthma whose onset can be traced to spinal injuries or concussions.

Given this reality of patients with asthma, how can I in good conscience uniformly lump them into one or two diagnostic categories with their designated code numbers and then administer preset treatments that correspond to those arbitrarily fabricated labels? The standard response to this question from those steeped in conventional medical tradition is that I should follow this protocol because it has been proven to work. I would argue, however, that it works only in the most superficial and temporary sort of ways. This goes to the heart of the critique that holistic medicine makes regarding mainstream medical methodology. When it focuses on generalities and fails to take into account whole persons, the consequences of its interventions tend to be local and short lived. In spite of perceived local benefits, there are often negative repercussions to treatment that can manifest as new health problems. Medicine's preoccupation with local results makes it possible to ignore the adverse effects that a treatment can have on the person as a whole.

The very nature of conventional diagnosis requires that it *de-personalize* individual cases of illness such that they can no longer be differentiated from any other case of the "same" illness. Personal, unique, and individualizing characteristics are ignored because they are considered irrelevant to the task at hand, which is to discover signs and symptoms that confirm the suspected diagnostic entity. It is not at all surprising that patients would complain of feeling alienated from a medical system that deliberately chooses to ignore the personal differences that go along with their illnesses and the unique stories that could potentially provide the keys to their return to health. The price paid for diagnostic conformity can be the neglect of personal meaning and the purposes of illness in the larger context. Both can be prerequisite ingredients to a genuine and successful holistic healing process.

Capitulation to conformity is further reinforced by medicine's overwhelming reliance upon laboratory testing. As a consequence, patients often feel themselves reduced to the proverbial numbers on a chart. Lab values can become the exclusive focus of a physician's interest, while communication with patients may be maintained not because it helps the doctor's decision making process as much as it projects an image of congenial bedside manner.

Like diagnostic code numbers, lab values serve to instill a sense of order upon patients whose illnesses may be unpredictable, uncooperative, and anything but the stereotypes of the diagnoses that they have been assigned. Lab values also serve to quantify the otherwise qualitative experiences of the suffering individuals in question. Numerical lab values also give a false sense of control and the illusory impression that the manipulation of numbers is tantamount to the curing of disease. The numbers represented by lab values can give patient and physician alike the idea that treatment is a simple matter of getting the numbers to cooperate—to adhere and conform to the norm.

Getting the lab values to translate into improved health is a lot harder to accomplish than one would think. It is not uncommon, for example, for the anemic patient whose blood count improves with iron supplementation to still complain of chronic fatigue, or the person whose lipid profile turns out to be optimal after a course of statin drug therapy to subsequently have a heart attack. Manipulating lab values cannot be equated with healing. It should go without saying that the numbers do not necessarily reflect the reality of patients, their circumstances, or their illnesses.

Lab values represent medical science's attempt to quantify illness. They give the impression of order, predictability, and control over illness. As such, they are frequently given far too much attention. When there is no therapeutic option left in a given situation, sometimes the only remaining course of action is to periodically recheck the lab values. A great deal of medicine involves the monitoring of conditions for which there are no treatments.

It is somewhat ironic, then, when we consider that the upper and lower limits of "normal" of any given range of medical lab values are artificial limits set by medical scientists who are appointed to determine such things. Normal ranges for lab values are mathematical approximations arrived at by a consensus of scientists. In other words, there is no actual hard line above or below which a lab value becomes normal or abnormal—there are only upper and lower limits of normal as determined by statistical probability and agreed upon by a group of earnest, yet fallible human beings. Normal lab values are not independent truths—they are man-made endpoints that facilitate the objectification of disease processes.

One rather obvious example of this is the manner in which medicine periodically takes the liberty to move the goal posts, so to speak, for certain values. What was once considered the acceptable range for normal blood pressure has been changed and is no longer applicable to patients today. The same is true of lipid profile standards; the experts have tightened up the numbers several times in recent years. In both cases, the new values are used to justify the need for prescription drugs for an increasing number of individuals.

In actual practice, it is common for lab values that fall just outside the norm to be dismissed by physicians as inconsequential. This is especially the case when the abnormal value is believed to be of no relevance to the diagnosis at hand. An abnormal lab finding may also be ignored when its significance in terms of the patient's problems cannot be determined. It may be labeled an *artifact*, which is really a convenient and face-saving term that allows scientists to ignore anomalies.

Some patients with abnormal lab values feel just fine, while others with normal values can have a variety of complaints and may feel awful. In spite of first-hand experiential evidence to the contrary, a doctor may tell persons in the latter category that, according the lab work, there is "nothing wrong" with them. The status of an individual's health is not necessarily reflected in the numbers revealed by a lab workup. The numbers may provide useful clues, can be misleading when they point to a problem where there is none, or may give a false impression of normalcy.

Medicine gives the appearance of being scientific when it abstracts information from actual patients by converting their ailments into diagnostic codes and laboratory values. We have been conditioned to believe that these are the activities that constitute good medical science. After all, isn't medical science supposed to distill illness down to the hard cold quantifiable facts? Conventional medicine is one particular approach, but there are a number of alternative medical modalities that are equally scientific in their own rights. They involve, however, completely different—oftentimes qualitative—methodologies.

Although the act of quantifying physiological processes may confer an aura of scientific validity to conventional medical methodology, it does not necessarily translate into control, efficacy, or any degree of certainty when it comes to the final outcome of a particular medical intervention. What it does do is render the language of the patient and his or her illness into a language that is compatible with the quantitative methods and objectivist agenda of modern medicine.

Some of Western medicine's purposes are conscious, such as satisfying the demand for conformity to its quantitative methodology, and some are unconscious, such as providing the rationalizations for why subjective information is thought to be of little value in the scientific process. Medical science excludes subjective experience because it is much harder, if not impossible, to quantify. It crafts its methods in a self-fulfilling manner, to confirm its own bias in favor of quantifiable phenomena.

One of the more far-reaching consequences of medical science's quest for uniformity is the stranglehold that modern pharmaceutical companies have on the therapeutic options available to the public. When medicine begins with the premise that standardization is not only desirable but also emblematic of properly conducted science, then it must conclude that most "natural" products do not meet this requirement. Non-synthetic plant, animal, and mineral-based substances found in nature are, by definition, viewed as suspect because they are not homogenous. Nature is not predictable; no one medicinal plant is exactly the same as any other plant, even when they are of the same variety.

Universalism is another term that effectively illustrates medicine's excessive allegiance to homogeneity and predictability. Just as the same diagnostic criteria must be universally applicable to all persons suffering from a particular disease, and all persons with a particular diagnosis must therefore have the same fundamental health problem, so, too, all persons with the same disease must respond favorably to the same medicinal substances, all of which must be free from impurities and identical in character. This is a very problematic assumption to make regarding a phenomenon as diverse and complex as human health and illness. It should be self-evident that a uniform approach to healing would be fraught with difficulties from the get-go.

The only way to overcome the allegedly undesirable lack of homogeneity of nature is to synthesize medicinal substances under controlled conditions in sterile laboratories. In this way, "active ingredients" can be isolated, "impurities" can be removed, and quantities of desired chemical compounds can be assured. Of course, impurities and active ingredients are viewed as such according to the eyes of the beholder. A substance may be regarded as an impurity, for example, simply because its true function has never been determined.

Herbalists and other holistically minded persons argue that plant substances in their natural forms can, when properly used, convey more benefits than a single, concentrated, so-called active ingredient. No doubt, the overall synergistic effects of the innumerable elemental components and chemical compounds that comprise a plant in its natural form cannot and will likely never be understood from the limited perspective of reductionist science.

It is no coincidence, then, that since human illness is also part of nature, it is distinguished by the same lack of homogeneity. Medicine solves this problem in the same manner that manufacturing synthetic drugs solves the unpredictability of natural medicinal substances—it formulates artificially conceived diagnostic categories in order to create the illusion of consistency and control.

Modern medicine believes that its synthetic products are more reliable than their naturally occurring counterparts and are, therefore, superior to anything that Mother Nature has produced over eons of time. In spite of this obvious bias as seen from a holistic perspective, medical science continues to maintain this stance in deference to the principles of uniformity and conformity. When science's resources are spent on defending its methodology, the real casualties become creativity, receptivity, and the freedom to think in new ways.

We have arrived at an odd impasse whereby phenomena that are quantifiable, predictable, reproducible, and categorizable are appropriate subjects for scientific investigation, while phenomena that do not meet this litmus test are not. By this logic, the study of the composition of the Earth is believed to be more scientific than the study of psychic phenomena. We cannot see, hear, touch, quantify, predict, or reproduce dreams—or so it is thought. Therefore, dreams are not worthy of scientific attention. While geology is a legitimate field of scientific inquiry, the study of dreams is not only not a recognized branch of science, but the very essence and reality of dreams themselves are called into question.

Some hard-core scientific skeptics go so far as to dismiss dreams as mere figments of the imagination. In other words, since dreams do not conform to the criteria laid down by the rules of science, they cannot be considered real. The study of dreams constitutes pseudoscience and would be a waste of intellectual capital. Here again, we see how medical science is enslaved by its own ideology. By contrast, I frequently make effective use of the dreams that patients reveal to me in my own medical practice, both as diagnostic clues and as therapeutic indicators.

Such a rigidly defined and distorted concept of what science is supposed to be must inevitably lead to the death of innovation. Medicine's need for uniformity and predictability leaves little, if any, room for originality or change. To many, the medical sciences are well established and not expected to change in any substantive way. All that remains, it is believed, is to fill in the remaining small details—such as the structure and function of the human genome. As the details become clearer, medicine will finally begin to produce the breakthroughs that it has been promising for so long. All the details in the world, however, will be unable to disabuse medicine of the skewed perspective imposed by its presuppositional –*isms*.

To outside observers with a broader and more holistic perspective, those promised breakthroughs amount to a great deal of wishful thinking. Such promises are not realistic because they represent a limited understanding of human health that is not compatible with holistic reality. The true purpose of science, to learn about the world around and within us, has been replaced by an imposter that insists that phenomena worthy of investigation must first conform to its rules and its conception of what acceptable science is supposed to be.

Perhaps the most striking example of medical dedication to a dysfunctional methodology can be seen in the way that research trials are conceived, conducted, and revered by the profession. I will have more to say about this later but, for now, I will simply point out how medical research, as fickle, contradictory, and corrupt as it is, nevertheless remains the authoritative source of evidence that justifies the standards of practice to which the profession is expected to conform. When physicians deviate from protocol, there is always some study that can be held over their heads in order to cow them back into submission. Unfortunately, the so-called "gold standard" of medical research is, in actuality, an emperor with no clothes.

The sterile homogeneity of conventional medical theory and practice is commonly and mistakenly believed to be the natural consequence of correctly applied scientific method. It is my belief, however, that the desire for uniformity is a built-in bias of establishment medicine that results in a distorted misuse of scientific method. Reproducibility is not and should not be an inherent characteristic of scientific method. If one starts with the premise that the primary objective of medicine is to look for commonalities among diseases in order to apply uniform treatments, then that will be the result. When the profession holds this methodology up as an ideal, it constructs a medical system that reflects that notion, regardless of whether it corresponds to the reality of patients and their complaints, and regardless of the disappointing clinical outcomes that it produces.

Medicine simply assumes that there is one best way to handle an illness, and that way—its way—will always come out on top, trumping all other therapeutic methods. A truly objective study of the beliefs and practices of Western medicine could easily lead one to believe that conformism is a fundamental feature of science. I believe that the real purpose of conformism in medical science is to project a sense of understanding and control in the face of the unsettling mysteries of human illness. It is a self-protective psychological mechanism that shields the medical ego from the reality of the shortcomings of its ideologically constrained methods of delivering health care. It also represents a subconscious agenda injected into science that creates the illusion of an objective universe compatible with the *-isms* of conventional medicine.

When medicine falls prey to a groupthink zeitgeist, the myriad subjective anomalies that normally present themselves to observant clinicians become easier to dismiss as not relevant to the scientific process. They are presumed to fall outside the domain of medical science and allowed to drift quietly away.

The personal toll inflicted on patients subjected to this type of medical system should be of greatest concern to all involved. Those subjective factors deemed by medicine to be unscientific and of no importance to its goals are often the very same factors that make the critical difference in whether a person manages to find the personal strength to persevere over the course of an illness. Attitudes and practices that contribute to the dehumanization and depersonalization of patients are not merely insulting—they often contribute to morbidity and mortality.

Methods that demand conformity exclude the unique and individual differences that go to the very core of who we are as human beings. Medicine, as it is currently conceived, generalizes in order to be broadly applicable to larger populations in the name of efficiency and profitability. This places medical theory in direct conflict with the inherently personal and diverse nature of human illness.

When we believe that people's ailments correspond to prefabricated categories, we also naturally tend to apply stereotyped treatments to those conditions—hence, our current cookie-cutter medical system, whereby one size is expected to fit all. The result is a large number of unhappy patients for whom the "indicated" therapy did not produce the desired outcome. This should come as no surprise since human illness is multifactorial and far too complex to be standardized, homogenized, and made to conform to diagnostic stereotypes. Regardless of how Western medicine wishes to portray itself, the study of health and healing is far from a hard and fast science.

One cannot generalize from one person's illness to another except in the most superficial of ways. The success of a therapy should not be judged on the breadth and ease of its applicability. Individualized therapies, which by their very nature apply to specific persons and their unique illnesses, are no less scientific than more broadly applicable therapies. Holistic reality demands that the physician regard each and every patient as he or she is, unencumbered from preconceived notions of what their illnesses are supposed to be like. Only then will the most appropriate treatments find their way to the right persons.

CHAPTER 9

Medical Dualism

The first peace, which is the most important, is that which comes within the souls of people when they realize their relationship, their oneness, with the universe and all its powers.
    –Black Elk, *The Sacred Pipe*

The yin-yang principle is not, therefore, what we would ordinarily call a dualism, but rather an explicit duality expressing an implicit unity.
    –Alan Watts, *Tao: The Watercourse Way*

D ualism is a philosophy of divide and conquer, where opposites do battle and the superior position prevails. Dualism is the first reductive step that divides existence in two. It creates an illusory world of alternative choices—black versus white, left versus right, dark versus light. It splits the world into polar opposites and treats them in isolation, as if one has little or nothing to do with the other. It rarely takes into account that paired opposites are almost always two sides of the same coin.

Dualistic thinking is so pervasive in Western cultures that it is presumed by the vast majority to be the only way of approaching the many situations encountered in life. The belief that one can disassemble, dissect, or subject to magnification in order to examine an issue, a subject, or thing in isolation and then understand it apart from the whole is the very essence of dualism. It leads us to believe that parts contain the secrets, the key to understanding. It is both a function of modern civilization's mode of thinking and the source of much of the Western world's dysfunction.

A dualistic perspective is the product of left-brain-dominant thinking; the kind of thinking that defines the scientific worldview. The left-brain mode of perception is an analytical mode that functions primarily by differentiating one thing from another. It uses reason and logic to compare and contrast. It views the world as a series of distinctions, differences, and juxtapositions. It sets people, things, and qualities at odds by emphasizing their dissimilarities. One can say without reservation that this dualistic scientific worldview *is*, in essence, the Western worldview.

Western hubris would have us believe that its dualistic take on reality is an inherently superior and fundamental feature of what it means to be human. While it is the defining mode of Western thought, it is not an inevitability. And although dualism is evident even in Eastern cultures, it has not historically been the predominant force behind Eastern thought. Granted, there are individual types of persons, natural right and left-brain thinkers, but Western societies are inordinately influenced by left-brain thought while, at the same time, undervaluing right-brain input.

Dualism is first and foremost about separation and division. We in the West are presented with a multitude of conflicting dichotomies. The polarization of Western culture has reached its raging peak in our own twenty first century. We think in terms of liberalism and conservativism, evolution or creationism, science against religion, and democracy versus socialism, with little room for any gray area in between. Scientism and religious fundamentalism are extremist positions cut from the same dualistic cloth. The sad fact is that many see these alternatives as exclusive, all or nothing options. People take sides and do battle with a winner-takes-all belief that only one side should prevail. This false choice between good and bad, right or wrong, superior and inferior, has become the lens through which we see the world.

Dualism is the very first *–ism*, the most basic presupposition that informs the scientific perspective. Dualism is the engine that drives all of the other *–isms*. Thanks to dualism we have empiricism versus rationalism, subjectivism versus objectivism, and reductionism versus holism. It also sets the stage in medicine for conflicts between materialism and vitalism, mechanism and energetics, and conformism versus plurality.

As such, dualism is an *a priori* ground rule of science, a philosophical assumption made by scientists, consciously or not, who take the material world full of perceived divisions as the one most important reality. This interpretation of reality comes down to a question of metaphysical definition and not, as science would have it, a matter of empirical proof. It is an ontological question regarding the nature of being. Is the world of separation and division created by science real, or just a consequence of science's methodology for examining that world? Are scientific classifications, subdivisions, diagnostic categories, and lines of demarcation real, or just convenient vehicles that meet the needs of scientific and medical paradigms? Is the real world a reductionist one full of myriad pieces, parts, factions, and features as Western science would have it, or is it a unified whole acting as one, as Eastern thought and indigenous cultures would have it? The answers to these metaphysical questions will inevitably determine the nature of the science and medicine that are practiced.

The list of polar opposites birthed via left-brain thought is almost endless. Dualism separates natural science from supernatural studies. It also conveniently separates science from philosophy, metaphysics, and religion. As a result, it exempts science from the scrutiny that it deserves because, we are told, science and values are separate, unrelated issues. Consequently, merely invoking the word *science* is usually sufficient to justify the actions of science. The same lack of accountability applies to the term *medicine*, which is assumed by many to be legitimate simply because it is medicine.

Continuing with our list of science and medicine-related dichotomies, we have positive and negative, protons and electrons, particles and waves, Newtonian and quantum physics, material and immaterial, the part and the whole, quantitative and qualitative, certainty and mystery, abstraction and observation, left-brain and right, astronomy and astrology, facts and anecdotes, hard science and soft, real science and pseudoscience, specialists and generalists, synthetic and organic, and theory and experience—to name just a few.

Perhaps the most famous of all dichotomies is the mind versus matter controversy, also known in philosophy of science and philosophy of mind as the *mind-body problem*. What this amounts to, in short, is the question of how it is possible for the mind to influence the body, and vice versa, when there is no discernable mechanism by which this relationship can be explained. There is no satisfactory scientific mechanism of action that allows for mind to have an impact upon matter. A good number of prominent scientists and great philosophical minds have attempted to resolve this dilemma, only to come up empty handed. This has given rise to a variety of proposed solutions, each of which is colored by the beliefs that underpin the particular solution. Not surprisingly, the scientific view of the mind-body problem is defined by its need for a mechanistic solution.

Like all good philosophers and scientists, we begin by defining our terms. The body is the seemingly easy part of the equation. Body is the physical aspect of human life. It is the focus of conventional medicine in health and illness. The body is the objective material counterpart of the subjective mind. It is tangible, palpable, and therefore measurable and quantifiable. The physical organ most closely associated with mind is the brain and, to a much lesser extent, the spinal cord, organs of sensual perception, and the nervous system in general. Brain study is the domain of neuroscientists, some of whom are convinced that their profession will one day resolve the mind-body problem.

Mind is a much trickier thing to define. Mind includes a dizzying array of non-physical phenomena including thought, emotion, intuition, sensations, imagination, extra-sensory perception, desire, will, and intention. Mind includes both conscious and non-conscious aspects. As per Jungian theory, there is a good bit of evidence that points to both a personal and collective unconscious, the non-conscious repositories from which various phenomena including dreams are generated. Mental events are subjective events, discernable only to the individuals experiencing them. It is reasonable to argue that one's individual reality—all that is experienced—is mediated through the mind.

I do not wish to be sidetracked here, so, for convenience sake, I will use the terms *mind* and *consciousness* interchangeably. There are those who would disagree but it depends on how these terms are defined. Some contend that mind is a subset of the much broader spectrum of consciousness, and others say that they are one and the same.

An entire academic discipline, *philosophy of mind*, is dedicated to understanding the elusive and mysterious nature of consciousness but it, unfortunately, suffers from many of the same materialist limitations of thinking that constrain conventional science. The much broader field of *consciousness studies* provides more fertile ground for exploration of issues involving mind, mind-body, and consciousness. Consciousness studies is more inclusive because it is not hung up on providing a material brain-related answer to the problem of consciousness and because it does not discount the field of parapsychology and the reality of psychic phenomena.

One longstanding theory is that mind and matter are two different types of things. It is an unmistakably dualistic theory that acknowledges body and mind as distinct and separate. One is physical, the other non-physical; neither can be reduced to the other. Spirit and soul are properties of the non-physical realm. Matter is presumed to be the fundamental reality of the empirical world. This particular version of dualism is known as *substance dualism*, wherein material substance and spiritual substance are taken to be two independent types of being.

Dualism has its origins in Western thought with Plato, Aristotle and, later, Descartes. It has been the prevailing belief in the West for the past two thousand years. For all practical purposes, regardless of what some would claim, it continues to be the operative belief underpinning science and medicine to this day. It is interesting to note that although substance dualism does acknowledge the reality of the immaterial it, nevertheless, refers to it as a substance. The material bias in such nomenclature is clear.

Substance dualism puts Western science in a bind because it implies that there actually is an immaterial reality—one comprised of mind, consciousness, soul, and/or spirit. Science and medicine balk at such a notion and, so, contemporary philosophers have felt compelled to come up with other theories that are more compatible with the modern materialist perspective.

One such attempt to make the immaterial more palatable to the mind of science is called *property dualism*. Just like empiricism was redefined to include only the so-called objective dimension of experiential reality, property dualism similarly proposes that mind and body do exist as ontologically unique properties, however, they are believed to be properties of one single, material substance. This rather cumbersome and contradictory line of logic is typical of the left-brain propensity for rationalization. It is nothing more than materialism in disguise.

The alternative to dualism is *monism*, which maintains that mind and body are essentially the same thing. The most popular version of monism in modern times is *physicalism*, which, not surprisingly, contends that mind can be reduced to physical properties and processes. It is a materialist version of monism. One subset of physicalism is known as *epiphenomenalism*. According to this theory, mind is a by-product or epiphenomenon of the brain. This represents yet another backhanded attempt to acknowledge the existence of consciousness while at the same time rendering it subservient to the material domain and the scientists that study it. Epiphenomenalism is particularly enticing to neuroscientists who are fond of the prospect of someday being able to "locate" consciousness within the brain and defining it in terms of brain anatomy and function.

Another, more subtle variation of material monism is called *emergentism*. This theory gives a somewhat higher standing to consciousness as an emergent phenomenon that cannot be reduced back down to the material elements from which it originally emerged. Somehow, mind emerges from matter as a unique new quality that cannot subsequently be explained in physical terms. Consciousness is unique but, in the end, remains dependent upon the "real" world of physical matter for its existence. Emergentism, thus, becomes just another rationalization designed to maintain the independent sovereignty of a material worldview.

The dilemma of how to explain the missing link between brain and consciousness has puzzled contemporary philosophers and scientists for several decades now. In a paper published in 1983, Joseph Levine, an American professor of philosophy, introduced the term, "explanatory gap,"[23] to refer to the inability of the physiological sciences to account for psychological phenomena. A decade later, David Chalmers, an Australian academic, spoke of "the hard problem"[24] in bridging the gap between objective science and mental experience.

Such characterizations of the mind-body problem prompt a number of important questions. Is it even possible for reductionist science to explain the basis of subjective experience? For example, the phenomenon of pain can be described in neurophysiological terms, but it still does not account for the subjective experience of pain. How can physical properties explain immaterial phenomena? Why does science insist that this can be accomplished? Will advanced knowledge in neuroscience someday provide insight into the human emotional response to a beautiful work of art or an inspiring piece of music? Common sense would indicate not, but researchers and philosophers of a materialist persuasion still hold to the hope that an answer will present itself. Many assume it to be an inevitability, a simple matter of the current incompleteness of science, which will eventually achieve sufficient understanding, thereby providing the necessary details to fill in the "gaps."

It is worth noting that Professor Chalmers of "hard problem" fame has been referred to as a dualist who believes in the unique differences between physical matter and mental states, but he also believes in natural causation. In other words, he wishes to have it both ways—consciousness cannot be reduced to matter but will somehow, someday, be explained by matter. At bottom, he, too, is a materialist.

The inescapable irony is that such conclusions are essentially faith-based and have little to do with science. I personally have no problem with this type of conjecture, as long as it is not presented to the public as a conclusion drawn from scientific evidence.

The hard problem arises from an oxymoronic need to explain consciousness in material terms. Some have even gone as far as theorizing that consciousness is just another very complex form of matter. This amounts to another example of explaining away a problem by redefining terms. The hard problem is a hard problem only for materialists who remain hell-bent on providing mechanistic and/or reductionist explanations where there are none to be found.

Functionalism is yet another theory of consciousness, one that is favored primarily by cognitive psychologists. Here, the brain is compared to a computer while mind is the information processed by the brain. The brain is the hardware and mind is the software. According to this notion, mind is neither matter nor energy, but is still dependent on the physical brain for transmission of information. While this theory may be appealing to technological societies, it fails to account for consciousness other than in materialist principles. It represents psychologists trying to frame their subject matter as a harder science than it actually is.

These variations of monism—physicalism, epiphenomenalism, emergentism, and functionalism—originally arose as arguments against dualism. Each is designed to subsume consciousness under the banner of materialism, thus rendering physical existence the dominant or only reality. They are flawed theories in that they deny the true immaterial nature of consciousness and/or place consciousness in a position inferior to matter. It is no wonder that scientists and philosophers would find it a very hard problem indeed to simultaneously acknowledge a dualism of mind and body while, at the same time, proposing material monist solutions. Such solutions make little sense because they are contradictions in terms. By the way, there is an alternate theory of monism called *idealism*, which proposes that mind is all that exists. To seriously entertain such a notion would be to threaten the very foundations of science itself.

The field of psychiatry provides a good example of how material monist bias has shaped the profession to such an extent as to make the modality of psychotherapy seem almost unnecessary. Even though psychiatry has long been convinced of the organicity of mental diseases like depression, bipolar disorder, and schizophrenia, no diagnostic test or imaging technique has ever been developed to confirm this theory. Nevertheless, the material bias of medicine causes the profession to persist in this line of thinking in spite of voluminous information that points to a significant psychological component to these illnesses.

Thanks to material monism, Jung, Freud, and others—some of the great thinkers in Western history—have been relegated to mere footnotes in the annals of psychiatry. Biological psychiatry and drug treatment are now considered state-of-the-art. Science, once used as a method of investigation, has instead become an ideological agenda that seeks to reinforce its particular worldview, in this case, a material monist theory of mental illness. From such a perspective, psychotherapy becomes virtually pointless. In contrast, imagine for a moment what might happen if psychiatry were to adopt the polar opposite belief that all is mind.

The very existence of *stress* itself—that uniquely modern psychic strain that almost everyone recognizes and most have experienced at some time or other—would be called into question if not for the mind-body connection. Even medicine does not doubt the role of stress in disease development. But it offers only the most facile of bromides when it comes to stress management and spends very little time investigating the specifics of how various types of stress lead to different forms of illness, both mental and physical. Alas, stress is one of those intangibles that science cannot sink its teeth into, that will not bend to the metaphysical *–isms* of medicine.

I know it seems a bit confusing—because it is. Medical science is undeniably dualist but disingenuously tries to solve the mind-body problem by taking a seemingly monist position. Medicine segregates the subjective from the objective, the material from the immaterial, mind from matter and, in so doing, acknowledges the existence of these phenomena, only to then propose that the physical is the only thing worthy of its attention. This is more a form of denialism than it is material monism. The mind-body paradox will always remain unresolved when approached from a materialist perspective. It cannot be forced to conform to such a one-dimensional philosophy.

A large part of the problem stems from the fact that conventional science and medicine are hung up on requiring mechanisms to explain observed phenomena before they are willing to acknowledge the existence of such phenomena. However, as previously discussed, most causal explanations are, in actuality, only descriptions of the anatomical, physiological, and biochemical circumstances of disease—they are not true causes. Medicine is a victim of its own mechanist ideology. It refuses to confirm the existence of a variety of phenomena relating to human health simply because they cannot be explained in material terms. As a consequence, a great deal of holistic medical theory, practice, and evidence is dismissed because it makes no rational sense from a materialist perspective. One would think that unexplained phenomena would generate increased curiosity among the medical profession rather than conspicuous disinterest.

Physics is one of the few conventional scientific disciplines that has shed some light on the otherwise restricted worldview of Western dualism. Contributions from physics such as the wave/particle paradox of light, the uncertainty principle's realization of the impossibility of locating so-called particulate matter in atomic space, Einstein's discovery of mass-energy equivalence, and the non-local nature of interactions between objects separated by space, are just a few such ideas that point to an indivisible relationship between mind and matter, subject and object, and researchers and their research. Fritjof Capra helped popularize some of these ideas in his *Tao of Physics*, which points out the similarities between concepts regarding the holistic nature of the universe revealed by quantum mechanics and the ageless wisdom taught by Eastern spiritual disciplines.

Of course, physics merely scratches the surface when it acknowledges only the energetic phenomena detectable by the instruments of science. It does not take into account far subtler energies that are un-detectable to science but have been described by yogis, spiritual adepts, and psychics throughout the ages. It turns out that physics and Eastern wisdom both point to the same fundamental truth: dualism and sepa-ration are illusions, mere surface appearances set against the invisible background of an underlying unified field.

The net effect of both dualism and material monism is to exclude the most important dimension of human health—the human psyche—from medical science. It should be self evident that any philosophy of heal-ing that assumes that physical aspects of the human organism—cells, tissues, organs, organ systems—can be studied in isolation and treated locally as if they are not related to mind, spirit, and soul, and as if such treatment will have no appreciable impact upon the greater whole, is a deeply flawed philosophy. It requires only a modicum of common sense and some real world experience to recognize this. One would think that medical science would be willing to admit that which is al-ready obvious.

Dualism encourages the perpetuation of a dissociated form of medi-cal thinking. It leads to unnatural conclusions detached from the every-day realities of patients and their personal concerns. When a rheumatol-ogist prescribes drug therapy for an arthritic condition and the patient returns complaining of depression, a dualistic mindset makes it easy to view them as two distinct and unrelated problems, thereby justifying a referral to a psychiatrist for further evaluation of the "new" condi-tion. Mind—the vehicle by which we experience most things—cannot be swept under the rug simply because it does not make itself readily accessible to objective scientific method.

Dualism places blinders on patients and practitioners who are ac-cordingly convinced that their perceptions of small parts of the picture represent the entire picture. It is the same kind of thinking that con-vinces us of the ontological independence of side effects, which, we are to believe, are the unwanted and unintended effects of a drug, even though they are no different from the benefits of a drug. They are both effects of the same drug but we conveniently segregate the side effects from the desired ones. In similar fashion, synthetic drugs, the singular achievement of modern rational medicine, are automatically assumed to be superior to "natural" therapies, many of which are discovered through the empirical trial and error of experience. More importantly, dualism produces an incomplete understanding of human illness, which in turn leads to temporary palliation, a dangerous trend toward symptom suppression, and incomplete healing.

From this divided perspective, rationalism becomes a rival to empiricism, objectivism thumbs its nose at subjectivity, reductionism is pitted against holism, and materialism continues to labor under the delusion that it is the sole reality in a strictly physical universe. This brings us back again to the limitations of an excessively left-brain mode of thinking—its primary impulse is to divide the world into opposing camps. It is a function of the discriminatory nature of the rational mind, which seeks to compare, contrast, and identify differences, as opposed to the more holistic right-brain, which perceives patterns, makes connections, and is more adept at comprehending the big picture.

One potential response to the so-called mind-body dilemma is to throw consciousness out the window in order to preserve the purity of scientific method—this is what science has done for centuries. The other is to modify the methodology so as to integrate the ever-expanding anomaly in the room that increasingly threatens to derail current medical theory and practice. Not unlike Newtonian physics' relationship to quantum theory, conventional medical science must come to grips with the vast quantum world of medical dynamics, the likes of which it is incapable of understanding without a little help from other, less materialistic and more holistic sources of medical knowledge. Although holism is the polar opposite of conventional medical dualism, the two do not have to be at odds. There is a time and a place for both to work together, sometimes synergistically, and sometimes independently.

As we shall see, there are vast resources available to those interested in expanding their perspectives regarding medicine, health, and healing. This includes a wide variety of holistic disciplines and energy medicines such as acupuncture, Ayurvedic medicine, homeopathic medicine, Traditional Chinese medicine, Reiki, Shiatsu, reflexology, herbal medicine, and nutrition. It also includes shamanic healing and many varieties of indigenous medicine.

Other disciplines like yoga, meditation, prayer, chant, ritual, and rhythmic movement are not just spiritual practices but methods of maintaining the holistic integrity of body, mind, and spirit. Such disciplines, in the end, cannot help but contribute to health and well-being. These methods can be used not only to maintain health but also to heal states of ill health.

In terms of understanding consciousness, the enormous body of information provided by consciousness studies, paranormal studies, and the field of parapsychology takes us right up to that line where mind meets matter. As a conventional matter, it is that place where unexplained phenomena go to die from neglect. From a more enlightened perspective, it is the cutting edge where remarkable lessons can be learned—if only one is willing to open one's mind to the possibilities.

Some traditions capable of providing insight into consciousness are thousands of years old, like Taoism, Buddhism, and Vedanta. They teach us that there is a third way that transcends both dualism and monism. That way includes both, but is neither dualist nor monist. Although it sounds rather abstract, it is directly relevant to the medical profession in that it provides a perspective that would allow it to move beyond its current shortsighted approach to health and healing. It involves a new way of thinking that transcends the decidedly myopic –isms that prevent medicine from becoming the true art and science that it is intended to be. But none of this is possible unless we first open our minds to reevaluate the philosophical principles and metaphysical presuppositions that inform the practice of medicine.

Part II

Medical Claims to Knowledge

CHAPTER 10

Information, Knowledge & Abstraction

Though you pile fact upon fact until the heap of evidence seems to touch the sky, it is still nothing in comparison with totality, just as a distance of countless light-years still comes no closer to infinity than does a single centimeter.
　–Gai Eaton, *Science and the Myth of Progress*

Where is the Life we have lost in living? Where is the wisdom we have lost in knowledge? Where is the knowledge we have lost in information?
　–T. S. Eliot, *The Rock*

Medical Information

It is said that we live in the information age. Clearly, this moniker is closely allied to the computers and other communication and information devices that many of us use in our homes and of-fices. While most are comfortable living and navigating in this world of digital data, some are beginning to question the value of so much infor-mation. Information overload is an experience that we have all heard of and many can bear witness to.

Overload implies that there is a point at which a conclusion must be drawn, a decision made, or an action taken, since further information will be of little or no value in achieving that aim. More information does not necessarily lead to clarity or understanding. An overload of data can generate the opposite effect—it can obfuscate, confuse, and lead to greater indecision and even paralysis.

Many have fallen under the spell of the digital information age. They are seduced by the notion that it is just a matter of time until we obtain enough information to allow us to achieve significant breakthroughs in our understanding of the problems and mysteries of life—medical and otherwise. Only then, when we reach that critical mass of scientific information, will things become clear, thus enabling us to solve our problems of poverty, unemployment, ill health, addiction, alienation, and crime.

The information age is intimately associated with our love affair with technology. Undoubtedly, there are many benefits to technologies of all kinds but, just as data has become the goal instead of the means to something more important, so, too, technology tends to be invented and developed for its own sake with little regard for its impact upon the quality of our lives. Technological growth, nowadays, is largely a function of corporate greed and a few special interests like the military-industrial complex. It has less to do with the overall well-being of humankind.

It is also true that technologies create just as many and sometimes more problems than they have solved. The information age is moving so fast that few stop to contemplate and many are unable to comprehend the repercussions of so much data and technology. We have collectively suspended our judgment as a culture because we remain bewitched by the wonder and progress promised by the next technological development just around the corner.

Western medicine not only reflects this cultural trend but is held up as a model of technological excellence—it represents the cutting edge of practical application in the burgeoning world of biotechnological sciences. Medicine is the primary driving force behind many new technological developments. But medicine, in particular, suffers from information overload. Like bees in a buzzing hive, medical personnel are busily producing large quantities of that honey-like data, which serves first and foremost to sustain the viability of the medical colony. However, just as techniques used by beekeepers are designed to encourage an unnatural overproduction of honey, the medical establishment is so single-mindedly geared toward the production of data that it lacks the wherewithal when it comes to making practical sense of it all. We are long past medical information overload. In fact, the medical world is so preoccupied with data that it can no longer distinguish between quantity of information and quality of understanding.

If we stop to examine the nature and motivations of the runaway medical train, it is best to begin with some commonly used terms that seldom receive attention. As such, these terms are often mistakenly interchanged with little regard for the subtle differences in meanings that they should normally carry. One such term is *datum*. A datum is simply an isolated piece of information. It is a bit of data separate from all other bits of data. The femur is a bone in the leg. A membrane defines the boundary of a cell. Red cells are components of human blood. These are bits of data—each one is a datum.

The term *data* refers to a collection of bits of information that are known or assumed to be facts. It is the plural of datum. Various bits of data may or may not have any relation or connection to one another. The modern connotation of data also has a strong association with the quantitative. Medicine is fond of interpreting medical situations in terms of numerical quantities and, indeed, much medical jargon is borrowed from the terminology used in the field of probability and statistics. Thus we hear of data translated into survival rates, lab values, randomized trials, and statistical significance. This represents empirical data that has been abstracted into mathematical form.

Information is a term that means the facts obtained through investigation, instruction, or study. *Information* is commonly interchanged with *data* but the implication is that information is more organized and less random than data. However, there is no guarantee as to the factual accuracy of information—it may be dependable or questionable. Although information can refer to a random collection of unrelated bits of data, it is more typically a collection of details regarding a common area of knowledge. Red cells are components of blood. Red cells carry oxygen. White cells are also components of blood. White cells tend to collect at sites of infection. This is a body of information that tells us something about human blood. Each bit of this information is an empirical detail that has been repeatedly verified through observation and experimentation.

There is no question that medicine is the king of data. Much of that data has accumulated over time as a result of the empirical observations of basic research scientists. The quantitative contribution to that data is a more recent phenomenon. Once the basic components of blood were discovered it remained to quantify how many red cells and how many white cells there are in a healthy person, a person with an infection, a person with anemia, and so on. The zeal with which modern medicine regards its storehouse of data rivals only the fanaticism of some baseball fans who are known to be able to quote little known statistical facts pertaining to their favorite teams and players.

Medical Knowledge

The term *knowledge* suggests more than just a collection of data or body of information. Knowledge involves knowing, which entails some degree of certainty. One can possess a body of information or a collection of data the specifics of which may or may not be reliable and/or verifiable. A body of knowledge, on the other hand, implies established information that has been gained through education or first-hand experience. Knowledge is of a different order than that which we consider to be unsubstantiated opinion or pure belief. It is something that we believe to be justifiably true.

Knowledge of direct experience is first-hand knowledge. When I accidentally cut my finger with a steak knife I experience the blade cutting my finger and see the resulting wound and blood seeping from it. Second-hand knowledge is when I tell you about my experience with the steak knife. When you tell your friend about my experience it becomes third-hand knowledge, and so on. The finger laceration is an objective phenomenon to the extent that it is verifiable by someone else who can inspect the wound after it has been inflicted. However, the further we depart from first-hand knowledge the more likely it will become altered in translation and therefore less trustworthy. By the time a fifth person hears about my kitchen mishap it may be believed that I lost a pint of blood after being stabbed by a violent intruder.

Although conventional medicine chooses to ignore most subjective information, which it excludes from consideration on the ideological grounds that it is not objective, it is nevertheless quite possible to have first-hand knowledge of subjective experience. When I cut my finger I felt a sharp pain, which traveled into my hand and up to the wrist. Afterwards, there was a throbbing sensation in my finger. Only I can verify this type of first-hand subjective experience. Even though no one else can know it to be true, it does not make it any less true. Similarly, I can have first-hand experience of a dream or a mood. And you can have second-hand knowledge of my subjective experience when I tell you about my throbbing finger, dream, or mood. This information, too, is liable to become more uncertain the further it departs from the actual event. Note that my own subjective/objective experience of cutting my finger is more accurately characterized as first-hand *knowledge* while third or fourth-hand awareness of this event may be less certain and therefore characterized as *information*.

Medical discourse has devolved over the years into a truly odd state of affairs. It used to be the case that doctors would share and compare case histories of patients with each other. They would learn by discussing the various nuances of their clinical experiences and gain valuable tips from one another. While this may still be a preferred mode of communication among some doctors, it has gradually given way to the pressures of conformity and uniformity in the medical environment. Thus the notion that there can and should be differences in diagnosing and treating the same condition in different patients is a concept that gets little attention nowadays. Personal experience no longer seems to carry the clout that it once had.

The physician as a voice of clinical authority has been gradually displaced by the research studies of "reputable" medical journals published by "prestigious" organizations. Scientists and laypersons alike have come to the point where they tend to believe something that is printed on paper or stored in digital databases more than the first-hand observations and insights of the clinicians doing the actual work.

Journals, which once were filled with individual case histories, are now almost exclusively comprised of elaborate research studies involving large numbers of anonymous patients whom the authors have sometimes never even met. When we take into consideration that the patients involved in such studies do not know if they have received a placebo or the actual medicine under study, and that physicians administering the treatment usually do not know either, and that statisticians who have not met any of the patients may be the only ones who do know, then we are not really talking about normal clinical experience anymore.

This type of patient-doctor relationship—if it can be called a relationship at all—is highly unnatural. The data gleaned from such studies is more accurately characterized as third, fourth and even fifth-hand information. The statistical manipulation of raw data produced by such a study results in a final product that is expressed in terms of percentages, probabilities, and standard deviations from the so-called norm. As such, the results derived from those mathematical calculations can be considered yet another degree or two of separation from the actual first-hand experiences of patients and doctors.

Medical Abstraction

When the data pertaining to a given group of individuals includes only objective information that is derived from third, fourth, and fifth-hand sources, it loses its connection to those individuals. Knowledge of this sort also loses most of its meaning as it relates to the very real people who provided that information. And knowledge deliberately separated from meaning is the fundamental commodity produced by both modern medicine and science in general.

This process of abstraction—which begins with real patients, places them under artificial conditions in a research environment, and yields in the end an impersonal database of statistics for our consumption—is a process that strips all persons involved, including the clinicians, of their individuality. Knowledge produced in conformity with this clinically detached standard is another way in which the unspoken mandate for uniformity in the medical environment is met. Individual characteristics that might identify unique circumstances for given patients are whitewashed away through an unnatural process that is believed to yield superior information in the form of abstract and sterile quantitative analyses.

It is worth stopping for a moment to contemplate what we mean by the word *abstract*. When one abstracts, one considers something independently from other accompanying factors. Abstracting can also be defined as considering something as an idea as opposed to a concrete event. The more patients and their illnesses are abstracted, the more they are transformed into theoretical constructs. Think of an abstract painting, which is premeditatedly intended to be independent from concrete representation. Abstraction deals in theoretical ideals while giving less attention to actual, specific, concrete instances.

The process of abstraction transforms a particular instance, an actual event, into an idea that no longer has concrete existence. It removes the specific identifying features of an object or experience and yields a theory, thought, or notion of an abstract nature. Abstraction is a rational process that removes the unique characteristics that would allow us to discern the specific instance from which it was derived. It yields nonrepresentational information that teaches us little about the particular circumstance from which the abstraction arose. Let us take the abstract medical textbook definition of bronchitis as an example. One such textbook defines bronchitis as:

> *Inflammation of the mucous membranes of the bronchial airways, caused by irritation or infection, or both, by pathogen.*[25]

This is an abstracted version of bronchitis conceived by medical authorities intended to represent the stereotype of bronchitis. In order to meet the criteria of a definition that applies to all cases of bronchitis it must be stripped down to its barest bones. It must be framed in terms of the lowest common denominators that characterize bronchitis. In order to achieve this, other information that applies to some cases of bronchitis, but not all, must be removed from the definition. Thus we arrive at a conceptual mock-up that describes bronchitis as inflammation of the bronchial airways caused by a pathogen. By the way, note how this definition of bronchitis neatly conforms to reductionistic, mechanistic, and materialistic expectations. Bronchitis, therefore, is inflammation *found locally* in the bronchi, is *caused* by irritation or infection, and is attributed to a *material thing*, in this case, a pathogen.

Now let us compare this definition to some examples of actual patients who have received the diagnosis of bronchitis:

> *Person A* is a fifteen year old boy who spent three hours outside in ten degree weather exposed to cold winds while building a snowman. He subsequently developed a high fever, chills, and a very dry unproductive chest cough that is aggravated during the daytime and becomes noticeably improved while lying in bed at night.

> *Person B* is a forty-five year old woman who has a history of bronchitis, which only occurs after exposure to cold damp weather in the fall seasons. She never has a fever but always develops a wet cough that produces yellow expectoration. Her cough becomes worse whenever she tries to lie down, thus preventing her from getting a good night's sleep.

> *Person C* is a twenty-seven year old man whose father has just died in a sudden tragic car accident. He develops a persistent hacking cough caused by what he describes as a tickling sensation in his chest. Every time he begins to talk of his deceased father he wells up in tears and experiences a paroxysm of intense spasmodic coughing, which can only be calmed by sipping on an ice-cold drink.

Everyday common sense should cause one to debate whether these three cases, regardless of their common diagnosis, can really be considered the same condition. While it may be true that all three involve inflammation of the bronchial airways, that is about all that they have in common.

Now imagine that we are going to conduct a study involving a particular treatment for bronchitis and these three individuals, along with two hundred others, are enrolled in the study. Any person with first-hand clinical experience in handling bronchitis would have to agree that there is going to be a great deal of variation as to the presenting symptom patterns of the two-hundred cases. Given this variation, one has to wonder if such a narrowly defined concept of bronchitis serves any useful purpose other than to mislead us into believing that we are dealing with the same affliction in all of these individuals. This being the case, we should also be concerned about the reliability of the data produced by this kind of research.

A number of important questions arise when we carefully contemplate the nature of this type of medical trial. First, are we justified in calling all of these cases bronchitis? Is it reasonable to think, given the wide variation of individual presentations, that the majority of bronchitis cases should respond positively to the same treatment? Is it even desirable that they should respond to one or a few specific treatments? Does this type of diagnostic and therapeutic shortcut do justice to our patients? Why does medical science value a reductionist approach, which seeks to define bronchitis in such narrow terms, over other approaches? If we tried to limit a study to a particular version of bronchitis, as in the case of person B for example, how long would it take to find enough people matching this description to include in a study, and how feasible would that be? Could this be part of the reason why medical studies seem to be so fickle, yielding such varied results from one study to the next? I can pose more questions but, suffice it to say, there is more than enough reason to cause us to question the methods and reliability of contemporary medical research studies.

Medical information is further abstracted when it is measured, quantified, and transformed into lab values and diagnostic images. Technology in general tends to render first-hand experiences and natural phenomena into abstracted forms that can be more readily manipulated by the methods of science. A black and white x-ray image is a convenient way to view the interior landscape of the human body, but it is a weak substitute for an anatomical dissection that reveals the internal organs to the naked eye. A symbolic equation representing a chemical reaction is an extreme abstraction compared to an actual chemical process taking place inside the human body. A graph of a brain wave on a piece of paper tells us nothing about the thoughts passing through that brain.

I am puzzled by the fact that the quantitative manipulation of medical information is believed to be the epitome of objective methodology. A body of information cannot be considered objective simply because it represents a quantitative, statistical, or technological abstraction from empirical or experiential phenomena. While such information can certainly be called abstract, abstraction does not automatically make it objective.

CHAPTER 11

Description, Explanation & Speculation

… an abstraction is nothing else than the omission of part of the truth. The abstraction is well-founded when the conclusions drawn from it are not vitiated by the omitted truth.
–Alfred North Whitehead, *Modes of Thought*

Medical Description and Explanation

In its scientific beginnings, medicine was essentially descriptive in nature. The quantitative and technological innovations that encouraged abstraction came much later. After excluding all but the strictly material, medicine set its focus on describing the human body as accurately and as objectively as possible. Over a span of centuries it has successfully explored human anatomy in its most minute details. I can attest to this given the thousands of named structures that had to be memorized in my anatomy class many years ago. Medicine has also successfully and painstakingly described the physiological processes and biochemical events that take place within the human body. And although these fields may yet yield useful information, anatomy, physiology, and biochemistry are now well defined, fully established sciences.

A good deal of confusion arises, though, when we accept description as a substitute for explanation. Scientific information is best obtained via observation of phenomena in their natural states. Interpretation of that information, however, cannot be equated with the empirical information itself. One involves observation while the other involves drawing conclusions about that which has been observed. They are not the same. When medicine offers descriptions of human anatomy, physiology, and biochemistry as explanations for abnormal health conditions, it makes a crucial error that can lead to misunderstandings and ill-advised therapeutic strategies.

For example, when a patient asks why she has asthma, a doctor may explain that the respiratory airways become constricted, thus allowing less air to pass through. This is a description of a physiological mechanism, not an explanation for how or why this particular patient has contracted asthma. When another patient wonders why he has migraines, he is told that they are a result of blood vessels contracting and relaxing in the head. This too is a description, not an explanatory determination of causation.

A description of something *as it is* is not the same as explaining *why* it is and, yet, most people, including medical professionals, do not always differentiate between the two. When a concerned and curious patient inquires as to the reasons for his or her illness, the doctor often provides a description of the condition and its associated anatomical, physiological and/or biochemical details. By virtue of our collective indoctrination into conventional medical culture, we are inclined to uncritically accept such descriptions as satisfactory explanations. In a certain sense, this type of material description of the details of an illness provides a limited explanation as to *what* is going inside the body of a person who has a given illness, but it almost never explains *why* that person has the illness.

This relates back to our previous discussion of proximate cause and bears repeating. Due to orthodox medicine's self-imposed mandate to ignore all but the material aspects of human health, it must at some point in the investigation of an illness settle for an arbitrarily chosen endpoint that defines one of the physical parameters of that illness. Descriptions of illnesses as defined by those parameters may serve as explanations of their anatomical locations, physiological mechanisms, and biochemical underpinnings, but they are not explanations of causation as to why any given person has an illness. Medical science's limited conceptual framework prevents it from being able to identify any deeper cause beyond the proximate material cause of an illness.

Let us use gout as an example. We are told by medical science that gout is a painful type of arthritic/rheumatic condition caused by sharply pointed urate crystals that form especially in the joints of the knee or big toe. But why does this happen? When too much uric acid accumulates in the bloodstream it precipitates into urate crystals. But why does uric acid accumulate in the blood? When the body metabolizes high levels of purines it produces more uric acid than normal. Why does this happen? The kidneys may not be able to adequately remove and excrete the high levels of uric acid from the blood. And why does this happen? This can result from a poor diet and overindulgence in certain kinds of foods that contain high levels of purines.

Now, we have just described the anatomical nature of gout (sharp crystals in joints), the physiological process associated with gout (uric acid accumulation in the bloodstream), and the biochemical mechanism underlying gout (purine metabolism), but we really have not explained why gout happens. It may be an accurate description and it may offer a limited understanding of proximate causation in material terms but it does not explain primary or actual causation.

There are many more questions regarding gout that can be asked. Why doesn't Bill develop gout even though he eats the same purine-rich diet as Joe who does have gout? Why does Joe's gout affect his knee but Jane's involves her big toe? Is it possible that Joe's request for a job promotion, which was declined two months earlier, have anything to do with the onset of his gout? Was there a connection between the onset of gout and the death of Jane's father three weeks earlier? Marie has been taking a diuretic medication for years. Could it have contributed to the onset of her gout?

As you can see, the determination of ultimate causation is a complex issue. On the other hand, the assignment of proximate causation is really quite arbitrary and depends on how one perceives and describes a situation. If we stick to the standard explanation that gout is *caused* by urate crystals that inflame the involved joints, then an anti-inflammatory drug like colchicine might be considered a logical therapeutic option. But if colchicine must be taken indefinitely to prevent future gout attacks, then we must question whether it is really aimed at the root cause of gout. And when Joe carefully watches what he eats but continues to have gout attacks, we have to wonder if his diet is the root cause.

Most so-called causal medical *explanations* are really abstracted *descriptions* of the stereotypical phenomena associated with a particular medical condition. When proximate causation is assigned, proposed solutions are naturally based on that supposed cause. When causes are deduced on false grounds or based on incomplete information because only certain material factors can be taken into consideration, it is not reasonable to expect satisfactory results from treatments developed via such faulty logic.

The salient point here is that empirical medical description is not an acceptable substitute for explanation or causation. They are not interchangeable. When we treat proximate causes based on faulty premises the outcomes tend to be less than satisfactory, short lived, and sometimes harmful. This type of approach may very well alter a disease process at some arbitrary point in its development, but it can also set off an unforeseen chain of events that may lead to new problems and further complications.

In other words, while an antipyretic like Advil, Tylenol, or aspirin may lower fever and provide some comfort for a child with a viral illness, it can also prolong the duration of illness by blunting the immune response. In addition, it can tip the balance in the wrong direction, encouraging the illness to morph into bronchitis or pneumonia, for example. Medicine looks the other way, assuming such events represent the inevitable course of the illness, not owning up to the role played by superficial treatment. Treating symptoms without taking into account their overall purpose is a recipe for potential trouble.

We cannot know all of the ramifications of our specific interventions on the micro level because, by definition, a reductionist approach takes into account only a very small piece of the much larger picture. This is the unreliable therapeutic dilemma that conventional medicine creates when it stakes claim to objective truth based on a reductive methodology that excludes non-physical reality and demands causal explanation as a prerequisite for acceptance.

Medical Rationalization

As we have seen, not only is medical methodology reductionistic, materialistic, and mechanistic, it is also disproportionately rational. Many assume the use of the term *rational* to be an affirmative characterization that references one of the strengths of scientific method. Rational, in the sense that I am using it, means overly analytical, adhering strictly to logic, to the exclusion of other means of obtaining and processing information, such as that which comes from empirical observation and subjective personal experience. It is erroneously believed by some that an uncompromising reliance upon rationality alone yields results that are more objective and closer to factual truth.

Although there are those who would object to this particular view of rationality, it is, nevertheless, the most accurate way to characterize the methodology employed in medical research and medicine in general. This methodology gives a distorted and false impression of medical reality when compared to the world of actual people and their actual health problems. *Rationalizations* are more likely to be the result of such an excessively rational methodology. A rationalization is an attempt to justify something using logic even when it is not true. Modern medicine is riddled with statements of so-called fact that are intended to be taken for objective truth but which are really elaborate rationalizations.

Examples of common medical rationalizations include the reduction of fat intake as a means to lose weight and increased calcium supplementation as a strategy to reduce bone loss. Both sound perfectly logical, which is precisely my point. In each case, logic notwithstanding, the recommended course of action is not likely to yield the desired results. Current nutritional theory points to processed carbohydrates and sugars as the prime culprits in the obesity epidemic. And the ingestion of large quantities of calcium will not necessarily counteract the body's lack of ability to retain that calcium. Nevertheless, the medical establishment endorses these approaches not because they are grounded in sound science, but because they seem to make logical sense.

Using antacids for the treatment of heartburn—or as medicine has redefined it, reflux disease—sounds logical, but what it really amounts to is an oversimplified rationale. Employing an antacid in the case of too much acid is logically appealing but not necessarily reflective of the true nature of heartburn. Similarly, prescribing anti-inflammatory drugs for inflamed conditions like tendinitis, bursitis, or arthritis, is based on logic more than empirical evidence of success—unless one's expectations of success are very low. It is well known that antacids and anti-inflammatory medicines need to be taken on a prolonged basis, sometimes indefinitely. While the use of such drugs is backed by logic, their effectiveness is short lived, and is not supported by longer-term practical outcomes.

Drugs used to combat conditions—classes of drugs that begin with the prefix *anti*—are popular in the medical world. Antidepressants, anti-inflammatories, antacids, antispasmodics, antibiotics, antivirals, and so on, have impressive sounding names. The idea conveyed by those names is logically appealing, especially to a shortsighted way of thinking. Of course it's good to fight disease, isn't it? Employing drugs to vanquish symptoms sounds quite logical, but does it really achieve its aims? Does it really work or does it just sound good? Is this type of medical strategy based on science or is it just a reflection of medicine's tendency to accept rationalizations in lieu of solid evidence?

Evidence for the inadvisable nature of medical rationalization lies in the poor outcomes that result from such thinking. One need not look very far to find a long trail of failed drugs that have caused much harm in spite of the fact that they have survived the supposedly rigorous (and exorbitant) approval process required by the FDA. Many a drug is conceived after numerous twists and turns of faulty logic. A particular stereotype of a disease is studied and a plan of treatment is devised based on its supposed (proximate) cause. A drug is then synthesized because of its theoretical ability to alter, inhibit, or suppress a biochemical and/or physiological process at some point in a chain of events that are thought to lead to that illness.

The method sounds very logical, scientific, and compatible with rational thinking, but it is not sound methodology because it does not address actual patients and the true underlying causes of their maladies. A professor of pharmacology once confided to me his misgivings regarding the modern process of pharmacological discovery. In essence, he complained about the tendency of the industry to pursue preconceived notions of what a drug should be able to do, as described above. He knew that millions, even billions, could be spent on an *idea*, a concept of how a theoretical drug *should* work if it *could* be synthesized, only to have it fail to live up to its anticipated potential once the actual substance had been successfully manufactured.

It is not hard to see why this methodology would yield such poor results. It starts with a stereotyped notion of what a given disease entails and assumes that this can be applied to most, if not all, persons with that disease. The physiology and biochemistry of the disease are studied while the way the disease and its treatment affect other body systems is ignored. All other mental, emotional, and spiritual aspects and concomitants of the disease are also excluded from investigation. Some arbitrary point in the supposed chain of causation in the physiological and/or biochemical development of the disease process is then chosen to be the site of chemical intervention. It is often chosen not because it will yield the best therapeutic outcomes but because it is believed that chemists are capable of engineering a substance that meets the necessary specifications.

Medicine's reductionist methodology does not take into account the bigger picture because that would constitute too many variables and too much complexity for scientific method to handle. Scientific method works much better when it considers fewer variables. The more variables that are excluded, the further from holistic reality will be its conclusions and outcomes. It increases the likelihood of encountering unforeseen consequences that could not have been anticipated given the exclusion of potentially vital information. A methodology so poorly suited to human needs must, of necessity, generate unreliable, unpredictable, and even dangerous results. Thus, my pharmacologist colleague's dismay over a process that relies on such a convoluted and overly rationalized version of scientific method is understandable.

When the circumstances under scientific consideration are cherry-picked to conform to a preconceived method of investigation, we should expect conclusions that are compatible with the worldview of that particular methodology. Hence, orthodox medicine produces reductionistic, materialistic, mechanistic, and logic-based solutions to problems that it presumes to be of the very same nature. Conventional medical science produces biochemical solutions for physical beings as if those beings were not endowed with emotion, meaning, consciousness, spirit, and soul. There is no place for psychic factors, energetic phenomena, longer-term outcomes, or holistic principles in such a worldview.

Even when, with the best of intentions, medicine tries to mimic the medicinal properties of nature, it does so in a reductionist manner by attempting, for example, to extract the so-called "active ingredient" from a given plant. Once isolated, this ingredient is studied apart from the many other natural plant compounds that it normally interacts with, as if those relationships are inconsequential to the properties of the active ingredient. A synthetic facsimile of the active ingredient is manufactured, often in slightly different form than the naturally occurring substance. The altered synthetic chemical can make subtle or not so subtle differences in the medicinal properties of the final product, sometimes causing unanticipated side effects or interactions. This process serves PhRMA's agenda well since the natural plant is essentially useless to the industry because it cannot be exclusively owned. A synthetic chemical, on the other hand, can be patented.

Now let us compare the methods of modern medical science to the way medicinal substances were discovered and studied in the past. In their purest form, the healing properties of naturally occurring substances were observed in nature by all cultures throughout all ages. Individuals would note the behaviors of animals and insects, and the characteristics and effects of plants and minerals. Mishaps like accidental poisonings provided much useful information. The medicine men and women of indigenous cultures accumulated a vast trove of therapeutic insights by simply observing nature in action. With nature as a guide and by the trial and error of personal experience, this knowledge was gradually refined, passed on through word of mouth, and maintained by cultural tradition.

Although modern skeptics would have us believe that indigenous medicinal practices sprang from the unscientific and superstitious beliefs of primitive savages, this turns out to be one of those distortions promulgated in the name of the myth of scientific progress. In reality, a great deal of vital natural healing wisdom has been handed down through generations of time. Note that this process involved no detailed knowledge of physiology or biochemistry. It was an empirical approach that employed first-hand observation and experiential trial and error. It was a hands-on approach that often yielded direct feedback of an immediate holistic real-world nature. In this sense, traditional methods of discovery may have been different, but they were no less scientific than the methods of contemporary science.

By the time early scientific observers got their hands on indigenous medicinal substances a great deal of knowledge had already been accumulated regarding their therapeutic indications and usage. Perhaps the most immediate and useful contribution of early scientists, therefore, was the documentation and systematization of this information. Therapeutic knowledge previously available only through oral tradition now became more readily accessible to a wider audience. The empirical tradition of trial and error continued as early physicians would note the results, for better or worse, of various interventions on behalf of their patients.

With the advent of biology and chemistry, scientists naturally began to investigate the various components, subcomponents and, ultimately, microcomponents of medicinal herbs, plants, minerals, and other compounds. The inner mechanics of the human body became less mysterious as its anatomical structure was revealed through careful dissection. A great deal of descriptive information regarding anatomy and physiology was gathered together. It was believed that the secrets to healing human illness would eventually be revealed through the study of individual body parts in isolation from the whole.

Medical Speculation

The next step was to make educated guesses about the function and relationships of various anatomical parts, and the nature of medicinal substances and their effects on the body. Here is where the scientific process became susceptible to rational speculation. In contrast to modern times, where the debate regarding the origins and treatment of disease gives the impression of being a settled matter, a wide variety of competing theories were put forth during the early phases of the budding new medical science.

Speculation abounded, and although it was restricted to material theories of illness, it was no less speculative than the supernatural theories of disease of the day. Medical history usually gives credit to Hippocrates for having rid medicine of superstition. In place of supernatural causes for disease, he proposed his theory of the four humors. He believed that disease was caused by an imbalance of the vital humors. Illness became a secular matter and has remained that way ever since. Another early theory held that illness was caused by miasmas, or noxious vapors. This turned out to be a precursor to germ theory, which continues to be the dominant school of thought to this day.

Conventional medical history tends to whitewash medical theories that suggest immaterial concepts of causation because they are mistakenly believed by mainstream medicine to have been disproven. Thus, history rarely speaks of disease as a blockage of chi as the Chinese would contend, or an imbalance of the vital force as homeopathy believes, or living in disharmony with one's environment as Ayurvedic theory explains. On the off chance that they are mentioned in medical history texts, they are given as examples of outdated superstitious thinking.

Nowadays, the only theories of disease believed worthy of consideration are materialist theories.

Empirical evidence accumulated from thousands of years of experience began to fall out of favor as the new theories of educated men of science gained increasing popularity. But the newly acquired authority of scientific medicine was based on a false belief that the profession had made significant strides in its understanding of the nature of disease and its cure. It had made strides, but mostly in non-holistic terms. Descriptive knowledge of disease was an important step, but it did not translate into knowledge of disease causes or their successful treatment. The results produced from equating description with explanation often turned out to be short-lived, palliative, suppressive, or injurious.

What had actually taken place was a rapid expansion in the quantity of information regarding anatomy, physiology, and the nature and composition of therapeutic agents. This increase in descriptive knowledge added to the quantity of data and provided insight into mechanisms by which biochemical, physiological, and anatomical processes take place, but it did not translate into a true holistic understanding of the mysterious process of disease development and its cure. In fact, as the scientific revolution began to pick up momentum, the very wisdom that *was* capable of supplying some of the answers ultimately came to be seen as outmoded, superstitious, and unworthy of the considerations of medical science. After all, that wisdom did not come from the scientific community, but from "uncivilized primitive" sources.

This largely continues to be the *modus operandi* of medicine today. Aside from a handful of important breakthroughs, little has changed over the past hundred years. While the database of descriptive information has swollen beyond imagination, there has been a corresponding diminishment of the requisite wisdom needed to understand that information. Data has become a convenient and deceptive substitute for the real knowledge that might otherwise offer a glimpse into why it is that people get sick and how to help them become genuinely well again.

The vast quantity of data, which no human mind could possibly absorb or retain, and which must therefore be stored in digital form, is one factor that has led to the contemporary phenomenon of medical special-ization. It allows the database to be split into more manageable subdo-mains of information. The average layperson is rendered mute in the face of such a daunting volume of information. It has duped everyone, including those of professional and political persuasion, into associating quantity of information with depth of comprehension.

What is most problematic, however, is the manner in which con-temporary medical science has taken the liberty of formulating elabo-rate theories of disease and treatment that have little basis in actuality. While such theories may constitute *logical* extensions of the descriptive data, they also tend to be arbitrary formulations grounded in a variety of erroneous suppositions. What was once considered the standard of reli-able medical knowledge has morphed from the collection of data based on empirical observation that it once was, into an array of speculative theories grounded in mechanistic, reductionistic, and materialistic bias.

Few seem to appreciate the significance of this turn toward faith-based speculative theorizing. The trend continues to masquerade un-der the rubric of objective scientific method, which, we are made to believe, is unbiased and beyond reproach. This is not dissimilar from the way a magician uses distraction and sleight of hand to maintain an illusion of uninterrupted continuity. This particular moment of scien-tific prestidigitation involves an unacknowledged shift from meticulous empirical observation toward an arbitrary form of conjecture. It is as if the popularity and successes of science have given medicine permission to indiscriminately speculate without having to answer to the rules of scientific evidence. Within mainstream medicine, merely claiming the mantle of science appears to be enough to deter most serious inquiries regarding the validity of its claims. By contrast, when some alternative medical modalities dare to claim the same, it can provoke a firestorm of ridicule.

And so, after observing an association between painfully inflamed joints and high uric acid levels, medical science leapt to the conclusion that gout is caused by too much uric acid. This is an example of re-ductionist bias as a larger disease process is reduced to a single microscopic component of the blood. The chemical imbalance explanation for depression is another example of magical scientific thinking. Once it was noted that some depressed individuals exhibit changes in serotonin levels in their brains, it seemed reasonable and logical to assign causation. Thus, the prevailing belief is that a serotonin imbalance is the root cause of mood changes in depressed individuals. Note the additional materialistic bias here in that the physical is presumed to cause the emotional. It is just as plausible, however, to claim the reverse in each case, in other words, that gout is a disease that produces high uric acid levels, and changes of mood can alter serotonin levels in the brain. It is no small point to make that if the causal explanations were reversed it *should* make an enormous difference in our therapeutic handling of the problem. In either case, the conclusions drawn are largely speculative and have little solid evidence to back them up.

Medical science wishes to have it both ways. It proudly claims to be faithful to a rigorous objective methodology while its theories and consequent actions are capricious and often contradictory to that claim. The source of this inconsistency lies at that moment when *descriptive details derived from empirical observation* magically transform into *concepts and explanations conjured up by logic and rationality.* Once logical statements are accepted as legitimate explanations, we become susceptible to any number of rationally persuasive arguments, each of which may be plausible but may or may not be grounded in *experiential* evidence borne out by positive patient outcomes. When logical explanations produced by analytical thinking are not grounded in personal experience or practical outcomes, they must be viewed as theoretical at best, and rationalizations at worst. Just because an idea is logical or rational does not necessarily mean that it is true, accurate, or even close to the mark.

There is a big difference between observed phenomena and speculative conclusions drawn from that information. Given the preponderance of rational liberties taken by medicine, and given the fact that they are often at variance with empirical findings, it is no wonder that the perceptions and experiences of the general public often clash with medical opinion, thereby creating a sense of distrust regarding the medical establishment. In this regard, the suspicions of the average scientifically uneducated person may not be unfounded—they often have a basis in reality. Their perspectives are not in sync with the purportedly objective world of rational medicine because they are formulated from other valid forms of knowledge that were long ago rejected by science. Those other legitimate forms of knowing derive from personal observations, ethical considerations, common sense, intuition, and first-hand experience. The conclusions drawn from personal impressions by laypersons not trained in the sciences can sometimes stand in stark contrast to the prevailing cultural myth that medical protocols are based in objective scientific fact.

Speculative theories should not normally pose much of a problem but for the fact that they often present themselves as scientific certainties. It is perfectly legitimate to hypothesize regarding some ailment and its potential treatment but, in the final analysis, it must be borne out in the experience of clinician and patient alike. Otherwise, it is untrustworthy and remains mere logic until satisfied patients are the beneficiaries of theory put into practice. Even then, short-term positive outcomes alone do not necessarily confirm the validity of a theory. The true litmus test is the longer-term well-being of the patient.

Unfortunately, much of modern medicine consists of this type of un-corroborated proposition predicated upon speculative logic. Most conventional medical treatments are justified on the basis of short-tem outcomes and do not provide certainty regarding anything beyond those specific instances. This has become increasingly clear as evidenced by the upward spiral of side effects, interactions, complications, medical errors, and adverse events that frequently outweigh the presumed benefits of some orthodox medical treatments.

The homeopathic method of diagnosis and treatment stands in sharp contrast to rational medical methodology because it deliberately avoids the natural inclination to explain why things are the way they are. It takes for granted that ultimate causation will always remain a mystery, and understands that all explanations are conditional. Practiced in its purest form, homeopathy relies almost exclusively upon a combination of empirical observations made by practitioners and first-hand descriptions of mental, emotional, and physical symptoms reported by patients. It takes patient reporting at face value and treats subjective information such as a fear of heights, feelings of jealousy, and a craving for spicy foods just as seriously as objective findings like elevated blood pressure and dry, cracking skin. Through a process of trial and error, symptom profiles are gathered, medicines that match those profiles are prescribed, and results are recorded at intervals of time. Progress is predicated upon the overall direction of the patient's health over time, including diminishment of symptoms, improvement of abnormal lab findings, and the patient's overall subjective sense of energy, vitality, and emotional well-being.

The substitution of logical argument for medical fact amounts to sophistry. Such arguments are made plausible only in the context of a reductionist worldview. It is easier to frame one's claims with conviction when only a small portion of the problem is taken into account. In this context, the false sense of certainty that accompanies logic and rationality constitutes a deceptive cover for the erroneous metaphysical assumptions that form the framework of scientific method.

When person A has a headache and drug B brings obvious relief, most physicians and patients are inclined to deem this a success. By extension, the mechanism of action of that drug and the physio-chemical pathway in the human body acted upon by that drug are then confirmed to be "true" because the drug is said to have "worked." The validity of the assumed physio-chemical causation and method of treatment of that headache are reinforced, thus bringing rational theory closer to accepted and sanctioned medical fact.

However, when we remove the reductionist blinders from our understanding of that same headache, we discover a number of disturbing details that must make us rethink our previous version of medical truth. It turns out, for example, that every time patient A experiences relief from a headache after taking drug B, he or she also has trouble sleeping for the next two or three nights. Patient A also notices the gradual development of afternoon sluggishness and irritability, which was not the case prior to beginning treatment with drug B. A medical workup reveals nothing of concern but patient A consents to taking an antidepressant after putting up with six months of sluggish irritability.

When we look past the artificially imposed constraints that forbid physicians from considering the bigger picture, surely we must conclude that something is amiss. The arbitrary application of logic to medical situations of this nature allows for the construction of convenient arguments of proof. It also creates opportunities for plausible deniability. And, so, it is suggested that there is no real verifiable connection between drug B and patient A's insomnia, sluggishness, and irritability. Why, those symptoms are pure coincidence, it is argued, and could have arisen due to any number of factors. We are told that until more research is conducted, patient A should continue taking drug B because there is no "conclusive evidence" that confirms a causative relationship between drug B and patient A's new complaints. In a clear reversal of the physician's dictum to do no harm, patents are advised to stay the course until harm can be proven.

It is relatively easy to construct speculative arguments to support one's viewpoint, especially when those arguments are divorced from the first-hand experiences of patients and doctors. In essence, logic, rationality, and the future outcomes of as yet to be conducted research studies—essentially a form of medical faith—take precedence over the experiences of actual patients. Patients are placed in the position of having to wait for the results of multi-million dollar research studies in order to have the conclusions drawn from their own personal observations and first-hand experiences validated.

If, on the other hand, doctors trusted their instincts and considered the possibility that there may be a connection between drug B and patient A's problems, then they would be compelled to reevaluate drug B's presumed mechanism of action and the alleged underlying cause of those headaches. The less obvious and more important question that is almost never posed is, if the headaches recur regularly and need continued treatment with drug B on a prolonged basis, and if the headaches return upon discontinuation of the drug, then how can this by any standard other than a reductionist point of view be considered a successful treatment?

Medical description is made possible by virtue of empirical observation. Medical description must never be confused with medical explanation. Some medical explanations are valid, some are true in a limited sense within the confines of the *–isms* of medicine, and many are ungrounded speculative theories that go unchallenged because they have the tacit approval of the scientific community. The scientific establishment's unreflective embrace of rationalism is the source of this unsettling trend. The application of logic to a problem does not automatically make it scientific, nor does it make it any more factual. Medicine must come to terms with this axiomatic truth and begin to reexamine some of its basic beliefs regarding the true nature of medical knowledge, health, and healing.

CHAPTER 12

Trials and Tribulations

A statistician is someone who tells you, when you've got your head in the fridge and your feet in the oven, that you are—on average—very comfortable.
    –Unknown author

The current established format used for clinical trials in medicine is a relatively recent phenomenon. It is said that the first modern research trial of its type was published in 1948 by epidemiologist, Austin Bradford Hill, who was investigating the treatment of tuberculosis with Streptomycin.[26] His was believed to be the first randomized clinical trial (RCT). Also known as randomized controlled trials, RCTs have become the widely accepted standard for modern medical research protocols. In fact, RCTs have been referred to as the standard for "rational therapeutics"[27] and are considered by the healthcare professions to be the "gold standard" for research trials.

What distinguishes RCTs is that they are randomized, blinded, and sometimes placebo-controlled. Study subjects are randomly allocated to receive one of two or more treatments or interventions, with the intention of removing the bias as to which subject receives which treatment or intervention. Many believe that the highest ideal is to randomly allocate test subjects to one group that receives the actual treatment while another group receives placebo. It is thought that a placebo-controlled trial allows direct comparisons to be made between the actual treatment and no treatment. Blinded studies prevent study subjects from knowing what treatment they have received. This is done to minimize bias in terms of subjects' assessment of their own progress, or lack thereof. In some studies, the researchers are also blinded in order to prevent bias in terms of interpreting results and to prevent favoring certain test subjects over others.

It should come as no surprise then, that the claimed strengths of randomized controlled trials are that they are consistent, thus meeting the desire for uniformity, and that they minimize bias, thereby remaining true to the principle of objectivity. RCTs are regarded by the medical profession to be the gold standard, the most reliable and bulletproof form of evidence produced by medical science today. RCTs supposedly stand at the pinnacle of the pyramid of medical evidence. The only form of evidence more highly prized are *meta-analyses* and *systematic reviews*, both of which involve the results gleaned from multiple RCTs.

Of lesser value, beneath RCTs in the pyramid of evidence, are other types of less rigorous research studies, followed by case studies reported by physicians. Lastly, at the bottom of the pyramid, we find the first-hand reports of patients, otherwise referred to as anecdotal evidence. Patient reports, therefore, are the medical equivalent of legal hearsay, and are considered the weakest and least desirable form of evidence.

This hierarchy of evidence forms the basis of what has come to be known as evidence-based medicine (EBM). The findings as determined by the pyramid of evidence of EBM, in turn, constitute the foundation of evidence-based medical practice. They are the guidelines by which well-informed physicians are expected to treat their patients. Thus, individual patients are expected to be treated in accordance with the evidence produced by oftentimes large research trials conducted on thousands of subjects.

This model has a powerful influence on determining what evidence is acceptable, what drugs are permitted to come to market, what drugs and surgical procedures are the preferred treatments, and what treatments are not to be trusted. EBM provides a compelling reason for physicians to acquiesce to a disturbing form of groupthink. As long as guidelines are followed, medical professionals can count on a certain degree of protection from ethical and professional liability. Deviate from the prevailing standards of practice, and one can expect potential professional and regulatory pressure from the medical establishment.

The practice of peer review serves to further reinforce conformity to the prevailing standards of medical practice. In order to be published, researchers' work must first be critically reviewed by a group of professional peers. Peer review ensures that research trials and their results are suitable for publication. Published papers must at least attempt to meet the criteria established by EBM. Peer review is designed to maintain standards and improve the quality of research. It also lends added credibility to published work because it supposedly meets the standards of the profession. In medicine, peer review can also apply to the practices of physicians whose clinical activities are evaluated by groups of physicians of similar training.

While the reasons for randomized controlled trial protocols, evidence-based medical practices, and peer review standards are well intentioned, such standards are, nevertheless, ripe for abuse and corruption. In reality, standards are set by groups of elite professionals who are in fundamental agreement as to what constitutes good medical science and good medicine. Since laypersons have little say in such matters, medical professionals are answerable only to themselves. They have free reign to set their own standards, acting as both judge and jury. The medical establishment decides which treatments are acceptable and the ones that are not, which papers get to be published, which journals and institutions are the most prestigious, who gets to be promoted, and who deserves to be reprimanded. Astonishingly, the industry is essentially left to police itself. Surely, this is an untenable set of circumstances for a profession that prides itself on its objectivity and freedom from bias.

An entire genre of books documenting the misconduct and profiteering of the medical-pharmaceutical-insurance industry has been published in recent years. I will, therefore, only make some brief remarks on the subject here. There are two broad issues in medical science and practice that I believe need to be understood more clearly, and they are bias and deception, whether consciously or unknowingly perpetrated. A surprising admission by Marcia Angell, M.D., former editor of one of the most prestigious American medical journals, speaks to the heart of the matter:

> *It is simply no longer possible to believe much of the clinical research that is published, or to rely on the judgment of trusted physicians or authoritative medical guidelines. I take no pleasure in this conclusion, which I reached slowly and reluctantly over my two decades as an editor of The New England Journal of Medicine.*[28]

Medicine has become so corrupted by corporate influence, vested interests, and tainted funding from PhRMA, big business, and government, that even some of its most dedicated servants no longer find it credible. Journal article authors routinely receive grant money, earn speaking fees, benefit from royalty monies, and sit on advisory boards for the very same companies whose products they are, in effect, promoting. Of course, not all researchers act in this manner but such practices are fast becoming the norm. Furthermore, these practices are commonly hidden from plain sight in the hope that they will not be noticed. Although authors are required to disclose conflicts of interest, they often go unreported and are sometimes even concealed through the employment of industry-sponsored ghostwriters—the actual authors whose connections to industry would delegitimize the research if their identities were to become known.

A related issue, which the medical profession does acknowledge, is that journals, editors, and researchers tend to exhibit bias against the publication of research results with negative findings regarding the product or practice under study. After all, it is rationalized, why would researchers want to report "failed" research that does not meet expected outcomes. Consequently, published research that consists of disproportionally positive outcomes gives a distorted impression of the sum total of actual research conducted.

In order to obtain FDA authorization, the drug approval process actually allows for trials that do not yield favorable results to be discarded in favor of trials that support desired outcomes, thereby falsely representing the available evidence. This should strike anyone with a sense of integrity as not only not objective but also unethical. It is common practice, sanctioned by FDA rules, for drug companies to disregard their own research studies that do not present new drugs in a positive light, and to continue to manipulate the variables of further research until the desired results can be achieved. The consequence of such deceptive practices is the bringing to market of products that are oftentimes later found to be ineffective and/or dangerous.

Some suggest that there should be no need for trials if a drug really does work. The benefits of a truly effective drug should be unmistakable, thereby rendering moot the need for elaborate studies to prove the obvious. Contemporary drug trials, it is argued, exist mainly for the purposes of getting dangerous drugs or drugs that do not work to the marketplace. We have seen evidence to confirm this many times in recent decades, with numerous products removed from circulation well after the harm that they have caused has been acknowledged. The *do no harm* dictum has been replaced by tacit permission for an industry to do whatever it can get away with until harm can be proven. No doubt you have heard that ubiquitous media advisory in the light of mounting evidence of an adverse drug effect to "consult your physician before discontinuing your medication."

Ongoing propaganda campaigns serve to twist the thinking of patients and medical professionals alike. In 2012, PhRMA spent 24 billion dollars alone on marketing to physicians.[29] This often takes the form of continuing medical education activities that allow physicians to earn CME credits, which are required by many states for physicians to maintain professional licensing. Such educational activities usually present industry products in the best possible light while downplaying side effects, dangers, and negative research findings.

The subterfuge and deceitful practices of the industry should come as no surprise and can be likened to the trickery employed by the tobacco industry, which once had the audacity to present its products to the general public as generally safe and benign. The use of *relative statistics* is a common way of distorting the evidence in order to give the appearance of positive benefits when there really are none. Amazingly, this misleading practice takes place out in the open while very few editorial boards or peer review panels seem to take issue with it. Allow me to explain.

Let's say that drug X is found to be helpful in 10 out of 1,000 study subjects. This translates into 1% of all patients receiving benefit. Now let's say that a new drug Z is studied alongside drug X and the results of the two are compared. It turns out that drug Z is found to be helpful in 15 out of 1,000 study subjects. In spite of the fact that this means that 1.5% of patients might benefit from drug Z, relative statistics are used to justify the conclusion that drug Z represents a 50% improvement over drug X. Technically, this is true because 15 is 50% more than 10. Ethically, this becomes highly questionable when reported as such to the medical profession and general public. It also represents very poor judgment on the part of professional journals that allow such findings to be framed in relative terms without clarification. Would you be more inclined to take a drug that is reported to be 50% better than another drug or would you prefer a drug that has a 1.5 % chance of helping your medical condition?

To further muddy the waters, technical jargon is often used to obfuscate actual results. Misleading terminology can give an impression of the success of a drug trial when in fact there is little to celebrate. We often hear that research trials yielded "significant" or "highly significant" results. What the average person does not know is that *significance* in this context is a measure of the likelihood of achieving such results *in comparison to random chance*. The results of a trial may be deemed significant if they are greater than that which would be found by mere chance.

I personally would need more convincing as to the benefits of a drug if those benefits were reported to be only somewhat better than chance. Significance does not usually translate into *practical* or *effective* in terms of real patients with real problems. It is another relative term that means something only in the context of a particular research trial. This is a kind of manipulative trickery that can fool professionals and even those charged with determining whether drugs should be allowed to market. The therapeutic benefits of new drugs are commonly inflated through the use of specious logic and false statistical advertising. This is clearly a case of the medical emperor wearing no clothes, where those involved are complicit in turning a blind eye to the greedy realities of the pharmaceutical marketplace.

Another misleading practice involves the manner in which doctors and researchers refer to *survival rates*. For example, when a person with lung cancer is told that he or she has a 50% survival rate, one would naturally assume that this means there is a 50% chance of beating the illness. What is often not mentioned is that survival rates refer to specified periods of time. Physicians have been known to be less than forthcoming when addressing their patients, preferring to call it a 50% survival rate rather than a 50% chance of still being alive after 5 years have elapsed.

It can be demoralizing when a false sense of optimism is conveyed by a doctor, only to have hope dashed when the patient learns afterwards of the true connotation of survival rates. When the media reports on survival rates published in journals, the medical profession is not inclined to clarify the true meaning of such terms, preferring instead to take advantage of the favorable press coverage. At bottom, this amounts to a misunderstanding that could be easily remedied if physicians were to be clear about the terminology that they employ but, sadly, this is not often the case.

Similar misunderstandings can arise around the use of the terms *effectiveness* and *efficacy*. We hear the word *efficacy* a lot in medical circles and it is easy to see why people would assume that it refers to how helpful a particular treatment can be for a particular ailment—but they would be mistaken. Efficacy actually refers to how well a given drug or intervention performed in the context of a clinical trial. It is another one of those relative terms that indicates how a drug performed in a research setting compared to another treatment or placebo. It says very little about how useful that intervention may be in helping sick patients.

Remember, an RCT may find that 10 subjects out of 1,000 received benefit from a flu drug, for example, but in real life terms that may mean that 1% of people with the flu who used that drug experienced a moderate reduction in the intensity of their aches and pains. It may be *efficacious* in reducing the sensation of achiness in a controlled trial, but it may not be very *effective* in reducing the time it takes to recover from the flu or in reducing the complications that can arise from flu. It could, in fact, turn out to prolong the duration of the flu and increase the likelihood of complications.

If RCTs that use relative statistics and relative terms to disguise the true meaning of their findings constitute the bedrock of proof used to justify evidence-based medicine, then we must seriously question the accuracy and reliability of the clinical guidelines produced by EBM. And this is exactly what is beginning to happen. The gold standard of evidence is becoming increasingly unpopular among mainstream physicians. A recent investigation by the *British Medical Journal (BMJ)* analyzed the effectiveness of treatments studied in 3,000 RCTs.[30] *BMJ* concluded that only 11% of treatments were found to be "beneficial," while another 23% were deemed "likely to be beneficial." Another 24% broke even in terms of "benefits and harms," while a whopping 50% were found to be of "unknown effectiveness." The remaining 8% were either "unlikely to be beneficial" or "likely to be ineffective or harmful." Given the biases that we have been discussing, even these numbers should be viewed with suspicion and are likely to be overly optimistic.

Among some of the more startling allegations are those of John P. A. Ioannidis, a physician and professor of medicine, whose technical paper, *Why Most Published Research Findings Are False*, has the distinction of being the most downloaded article ever published by the journal, *PLOS Medicine* (Public Library of Science). Without factoring in the overt economic corruption that pervades most of the industry, Ioannidis names some of the less obvious but very real biases that influence medicine:

Conflicts of interest are very common in biomedical research, and typically they are inadequately and sparsely reported. Prejudice may not necessarily have financial roots. Scientists in a given field may be prejudiced purely because of their belief in a scientific theory or commitment to their own findings. Many otherwise seemingly independent, university-based studies may be conducted for no other reason than to give physicians and researchers qualifications for promotion or tenure. Such nonfinancial conflicts may also lead to distorted reported results and interpretations. Prestigious investigators may suppress via the peer review process the appearance and dissemination of findings that refute their findings, thus condemning their field to perpetuate false dogma.[31]

When we consider corporate influences, personal human biases, deceptive language, statistical trickery, the downplaying of adverse events, the exclusion of evidence that does not support desired outcomes, self-policed peer review, and publication bias for and against favored and unfavored topics, individuals, and institutions, any reasonable person would be inclined to question the credibility of the self-anointed moniker, Evidence-Based Medicine. Like "peacekeeper missiles," EBM's claim to authority based upon the methodological purity of RCTs has a distinctly Orwellian ring to it. The modern ideal of clinical evidence appears to have little relevance to actual patients and their medical conditions.

Research trials from a holistic perspective

Most of what we have been discussing thus far represents a conventional critique of RCTs and EBM from a mainstream perspective. It does not take into account the holistic viewpoint, which raises additional questions about the relevance and applicability of EBM to real patients with real problems. RCTs are a product of materialist, reductionist, mechanist, rationalist, and objectivist bias. Most clinical trials are conducted without taking into account holistic principles. They rarely, if ever, pay heed to whole person risks and benefits, or the impact of conventional interventions upon heart, mind, and soul. Long-term consequences of this nature are often overlooked by clinicians because, if taken seriously, they would create a number of problems for the –isms of medicine.

It is important to understand that processes like disease development and genuine healing cannot be accurately or adequately represented by quantitative models based in numbers, averages, graphs, charts, and statistical probabilities. This is a self-evident truth that should require no further justification, at least to those who intuitively grasp the significance of holism as it relates to human health. I acknowledge that to those who view the issue from a predominantly left-brain perspective, this may not be so apparent.

By virtue of quantitative bias, outcomes that are measurable in more concrete terms—such as blood pressure readings, cholesterol profiles, bone density values, and so on—receive the greatest attention. Other less tangible factors such as motivation, will power, vitality, happiness, and general sense of well-being are rarely topics of scientific investigation. From a holistic perspective, these are some of the more important indicators of present and future health. This explains how research trials can yield positive results in terms of a drug's benefits, for example, even though a patient may experience a decline in health while taking that drug. By clinical trial standards the drug works but, by personal standards, it fails to result in improved overall health.

This examination of RCTs relates back to our discussion of abstraction. The further one departs from the first-hand experiential reality of actual patients, the less pertinent is the knowledge derived, regardless of the method used to obtain that information. A formulaic process that selectively pre-approves patients to participate in research trials, prevents them from knowing whether they are to receive one treatment or another or placebo, blinds attending doctors from knowing the same, uses statisticians who have had no contact with study subjects to compile and manipulate data, and ensures compliance with conventional standards via peer review, is a process so far removed from the experiences of individual patients as to be virtually irrelevant to their particular needs. RCTs do not produce first, second, third, or even fourth-hand information. RCTs represent the pinnacle of medical abstraction and, as such, have little of value to say about specific patients in real life and real time.

The most valued RCTs are ones that include large numbers of participants. Granted, the information derived from such RCTs may have some purpose when making facile generalizations about large populations—such as the likelihood that smoking may lead to lung cancer or excessive alcohol consumption can cause liver disease—but its value in specific situations for individuals with unique health issues is minimal at best. EBM assumes that conclusions regarding individuals can be made based on results derived from large population averages. In actuality, the larger the study, the less useful will be the results. Treatment guidelines for individual patients should not rely primarily upon the generalized type of information that is produced by EBM. By contrast, best practices can be achieved via a holistic approach that includes many sources of information, especially the first-hand information provided by patients themselves.

Some of the more fundamental assumptions behind RCTs come into direct conflict with tried and true holistic principles. Study subjects are chosen on the basis of stereotyped diagnostic criteria that conform to artificial conventional constructs of what diseases are supposed to be. Applicants that do not fit the litmus test are excluded in an attempt to eliminate disparities among participants. While this creates a comparatively homogeneous population of study subjects, it does not reflect the diversity represented by most individuals who carry that particular diagnosis. The quest for objectivity and uniformity results in the sacrifice of individual differences that are normally found in natural populations.

This one-size-fits-all mentality also presumes that all subjects in a study can and should respond favorably to the same therapeutic intervention being tested. This is a very important point that forms a sharp line of demarcation between conventional notions of treatment and more holistic approaches. Individualization of treatment tailored to individual persons with their unique health problems is one of the great hallmarks of holistic therapies. In holistic thinking it is taken for granted that one size does *not* fit all, and that this fundamental truth is more representative of actual patients with actual problems. If this really is the case, then imagine how difficult it would be to enroll a group of study subjects with the same conventional diagnosis who are essentially similar in all ways. It would be near impossible. Therefore, by definition, all groups of trial participants constitute artificial, unholistic, and unrealistic approximations of the general population at large.

In addition, RCTs are designed to account for only a limited number of variables. RCTs place a reductionist filter on the living systems that they study. They study organisms in parts, without context as to the greater whole. As the number of factors encompassed by a study rises, it becomes increasingly complex to manage to the point of becoming impracticable. Conventional medical science seems to be content with this narrowly focused approach to research.

RCTs are also unholistic in the sense that most are limited to the discovery of shorter-term results. Corporate sponsors are not interested in knowing the long-term consequences of their products. Their intent is to produce just enough data to deliver a drug to the marketplace. In similar fashion, conventional medical theory is, by its very nature, myopically focused on localized, short-term results. This is where the interests of corporations, RCTs, and orthodox medical theory dovetail. Holism, on the other hand, demands that any thorough evaluation must take into account the short and longer-term impact of all therapeutic interventions upon the whole person. To do otherwise is to fool oneself, to knowingly look the other way so as not to acknowledge the consequences of one's therapeutic shortsightedness.

RCTs place real people in highly artificial, sterile, and unrealistic circumstances that either ignore or completely bypass the role of the doctor-patient relationship. It is assumed that medicinal drug effects can be separated from all other factors, including human factors. Quantum physics taught us long ago that the mere act of observing an event can alter the nature of that event. Observer and observed are irrevocably interconnected, thus making any attempt to separate clinician from patient an artificial and futile exercise in false objectivity. All factors, for that matter, are holistically intertwined, making it impossible to fairly study the ailment of any patient in isolation from drugs, placebos, clinicians, clinical settings, socio-economic and educational factors, and all other health issues experienced by that patient. I am not arguing that medical research is without value. There is a reasonable middle ground between outright rejection of its methods and unbending faith in all that it represents.

Another defining feature of conventional RCTs is that they are designed and believed to be *reproducible*. We are all familiar with this aspect of scientific discovery that asserts that if the results of an experiment are to be believed then they must be capable of being replicated under the same or similar experimental conditions. Otherwise, the evidence produced by a single isolated study is not considered to be credible. And I would agree that this principle of reproducibility is true, but only up to a point. The more lifeless and inert the subject of investigation the more replicable the study results will tend to be. The boiling and freezing points of a liquid, for example, are phenomena that nicely satisfy the desire for reproducible results.

However, when we venture into the world of organic life forms, the subject matter becomes more complex, thereby making experimental results much more difficult to replicate. This applies especially to the most complex of all study subjects, human health. Not only will such results defy the dictates of scientific conformity; they should not be expected to be replicable. From a holistic perspective, it is not realistic and should not be considered a desirable goal. If we take into account issues of body, mind, and soul, it is not possible to find a group of individuals that are, in most if not all ways, of equivalent health status. The same disparate group cannot be expected to respond favorably to the same therapeutic interventions. If this inherent resistance to pigeonholing is true, then why must medical science insist otherwise? It does so only by bending holistic reality to suit its artificial worldview.

Finally, we must question the purported objectivity of an RCT approval process that is ultimately subject to review by a group of peers who, in essence, are chosen by virtue of their professional training and their agreement with fundamental assumptions about what constitutes health, illness, healing, and medicine. There is little evidence that peer review actually works—it is simply assumed to be the case.

Peer review is an influential process that can determine which research trials are funded, which papers get to see the light of day, and which scientists ultimately receive the recognition and approval of the scientific establishment. It is a costly process that creates a strong bias towards well-endowed scientific institutions and organizations. And it predisposes toward acceptance and publication of papers authored by persons affiliated with such organizations. The very same well-funded institutions are also the most highly regarded and prestigious ones. Not surprisingly, the institutions with the most capital, prestige, and influence are the ones whose medical authority carry the most weight.

It should go without saying that these institutions also represent establishment views regarding science and medicine. Peer review, rather than conferring objectivity to scientific activities, winds up functioning as an exclusive club that ensures conformity to majority opinion and beliefs. The same paper by Dr. Ioannidis that I cited earlier regarding the nature of research findings contains the following remarkable statement:

> *Claimed research findings may often be simply accurate measures of the prevailing bias.*[32]

I find this to be the most insightful statement in his paper. Perhaps the one clear and reliable thing that can be said about the entire RCT, peer review, EBM process is that it reinforces the built-in, *a priori* assumptions held by the conventional medical establishment. A very good case can be made for the idea that RCTs and EBM amount to an arbitrary litmus test imposed by modern medicine designed to keep competing ideas from receiving a fair hearing. They function largely to maintain the conventional medical status quo. Those wishing to publish research that does not reflect accepted medical theory, and which is not backed by a recognized institution, face a long uphill climb and are not likely to meet with success.

Those who practice complementary or alternative forms of medicine

(CAM) can testify to the knee-jerk cries of "where's the evidence?" when they encounter establishment critics. Many have been confronted with demands to cite RCTs that prove the efficacy of their particular CAM modality. But who in their right mind would expect to find favorable research in support of a CAM modality knowing full well establishment medicine's history of bias against CAM? When a positive research study regarding a CAM modality is cited, the typical response is that more research needs to be done before mainstream medicine is willing to take it seriously. This card tends to be played by skeptics regardless of the number of credible trials that are cited.

Such demands for proof put the cart before the horse. When I am made aware of an interesting idea that could potentially benefit my patients, my first inclination is to ask questions to get a better understanding. I do not automatically go on the offensive seeking to disprove that new idea. And yet this, in essence, is what EBM has become—a weapon used to defend establishment standards and to ward off challenges to orthodox dogma. Some who demand EBM findings to validate CAM modalities are being disingenuous. They have no real interest in scientific objectivity—that would require an open mind—because their true agenda is to discredit ideas that threaten their worldview. They cling to EBM like a fundamentalist clings to scripture.

Many CAM supporters are duped by such disingenuous cries for evidence and seek to appease the mainstream guardians of medical truth. They believe that if they can just compile sufficient evidence then their medical experiences will be vindicated. After years of hard work and millions of dollars spent they are likely to be told that it is still not enough. They fail to realize that the game is rigged. As long as the –isms of medicine hold sway and as long as PhRMA is guarding the henhouse, very few new ideas will be permitted entry. Although most who subscribe to EBM may genuinely believe that it safeguards state-of-the-art standards, subconsciously and in reality, it functions as a form of professional protectionism.

More to the point, RCTs and EBM have become tools to be manipulated in the game played between PhRMA and the FDA. Studies that do not support PhRMA's agenda are allowed to be discarded while corroborating trials are accepted by the FDA, which then looks the other way. A new drug is approved with flimsy evidence, the PR machine kicks in, and only after billions in profits have accrued does the general public begin to become aware of the shortcomings and dangers of that drug—until finally it is pulled from the market. You would be mistaken if you thought that such occurrences are simple oversights. They are a function of structural corruption built right into the medical system. Even when fined millions of dollars for defective or harmful products, PhRMA remains unperturbed and is not inclined to change the rules of the game. Such fines represent a small dent in drug company profit margins. In the end, the penalties for wrongdoing are simply written off as the cost of doing business.

Ironically, many widely practiced medical protocols are not supported by EBM. Even though they do not meet the medical profession's standards of proof they, nevertheless, elude critical review and are passed down from one medical generation to the next. Such practices are often justified by the fact that they can be found in authoritative textbooks and medical journals. I personally have no problem with these tradition-based medical practices, as long as they can be validated with clinical results and satisfied patients. However, it cannot be claimed that they meet up to the unique requirements of evidence-based medicine.

A recent Mayo Clinic review of ten years of medical literature revealed a high rate of what are referred to as medical reversals.[33] Medical reversals occur when newly published research contradicts or invalidates a currently existing practice or standard of care. The study concludes that reversals are common, thus highlighting the fact that many medical practices persist in spite of lack of evidence consistent with EBM.

The gold standard of medicine is badly tarnished. RCTs and EBM are not the arbiters of clinical truth that many believe them to be. As would be expected, persons in positions of authority set the standards and those standards reflect the unique biases of the medical establishment. Such standards do not reflect objectivity in the sense of an open-minded and receptive attitude toward innovative ideas that can be of value in helping sick patients. Medical science rejects most new ideas without giving them serious consideration, especially when it is found that they cannot be marketed or monetized. Homeopathic medicines, herbal medicines, nutritionals, and acupuncture needles are small potatoes when compared to blockbuster drugs. Mark Hyman, MD, a leading proponent of an alternative approach called functional medicine, sums it up nicely:

> What most physicians and consumers don't recognize is that science is now for sale; published data often misrepresents the truth, academic medical research has become corrupted by pharmaceutical money and special interests, and government regulators more often protect industry than the public. Increasingly, academic medical researchers are for hire, and research, once a pure activity of inquiry, is now a tool for promoting products.[34]

RCTs do not deserve their privileged position at the pinnacle of the evidentiary pyramid. Western culture has always placed inordinate value on novelty. RCTs just happen to be the new kid on the block. Simply because an elite class of professionals has ordained RCTs the highest form of medical evidence does not make it true. Medical science did just fine during the many decades prior to the rise of RCTs and EBM. The notion of RCTs as the epitome of medical proof is simply false. I do not deny that RCTs can play a positive role in medical discovery, but they are significantly overvalued when compared to all other forms of information and evidence.

Science is not the exclusive right of a privileged class of scientific elite to be practiced only by those with proper credentials affiliated with establishment institutions. The South American shaman who, through trial and error, uses plant-based medicines to heal the sick is also practicing a form of scientific discovery. The backyard gardener uses scientific method when he or she employs and compares a variety of methods to maximize yield. Scientific method is not a possession to be monopolized by a particular group, and it is most definitely not a brand or product to be marketed by those who claim ownership over it. Perhaps the most damning truth is that patient interest is just an afterthought when compared to the self-interested priorities of corporate medicine.

Admittedly, evidence-based medicine and holism have completely different agendas. They constitute different bodies of information derived from different perspectives. One focuses on short-term symptom elimination while the other concerns itself with whole person and long-term healing. Ideally, they should both play a role in any quality health care system. However, only one is bent on invalidating the other. One is fundamentally inclusive while the other is territorially exclusive. Nevertheless, there is no cogent reason why the two cannot work together.

There is also no reason why RCTs must represent the one and only source of medical truth. Holistic modalities cannot be dismissed simply because they are not compatible with the structural requirements of an RCT format. It would be like claiming that quantum physics should be rejected because it does not make sense according to the standards of Newtonian physics. This incompatibility is indicative of the limitations of RCTs and EBM. RCTs reinforce *a priori* assumptions that clash with holistic principles. Even though some see them as opposites, holism and reductionism are complementary perspectives. RCTs study parts of the picture while holism studies the complete picture. RCTs assume one treatment can be applicable to many, while holism believes that treatment must be tailored to the individual. When employed together, patients are the beneficiaries.

The knock against all other methods of gathering information is that they are less objective than RCTs—they are supposedly tainted by bias. While they may be susceptible to bias, the extent and danger of that bias is highly exaggerated. It is my contention that RCTs yield information that is least applicable to real patients precisely because they are infected by their own form of unconscious bias. The type of objectivity represented by RCTs is achieved only by decoupling research protocols from actual patient problems and doctor-patient relationships. The fact that

RCTs are the form of investigation most compatible with the many –*isms* of medicine also makes them the least relevant to the holistic reality of patients' lives.

It must be stated that despite my strong critique, medical research serves an important role in discovery and in expanding the knowledge base. It is not research itself that I object to as much as the deceptive practices employed by special interests and the unwarranted elevation of RCTs and EBM to their status as superior forms of medical evidence. While research will always be important, it must never be used to silence other sources of medical knowledge.

In my opinion, medical research has evolved in the direction of RCTs and EBM primarily because it allows scientists and clinicians to remain unaware of and detached from the shortcomings and dangers of pharmaceutical solutions. When attention is focused on the format—the procedural details and internal nuances of RCTs—it makes it easier to claim ignorance later on in terms of the real world implications of such trials. I believe this to be a sad but true reality. It is the very same reason why so much funding and professional energy is spent on diagnostic technologies—because the highly restrictive –*isms* of medicine severely limit the therapeutic options available to physicians. When treatment options are restricted, diagnosis becomes the default focus of scientific activity. Likewise, when the concerns of actual patients clash with the aims of medical objectivity, diverting one's attention to the theory and practice of properly conducted scientific methodology serves as a convenient distraction.

CHAPTER 13

Medical Reality

The storyteller makes no choice
Soon you will not hear his voice
His job is to shed light
Not to master
   –Robert Hunter, *Terrapin Station*

Certainty is the greatest of all illusions: whatever kind of fundamentalism it may underwrite, that of religion or of science, it is what the ancients meant by hubris. The only certainty, it seems to me, is that those who believe they are certainly right are certainly wrong.
   –Iain McGilchrist, *The Master and his Emissary*

For each of us as human subjects, what is more real than our joys and sorrows, hopes and fears, desires and beliefs, and our sensory experience of the world about us? On what grounds are we to believe that these mental phenomena are any less real than such physical phenomena as mountains and buildings, let alone quarks and electromagnetic fields?
   –B. Alan Wallace, *The Taboo of Subjectivity*

The practice of science is a collective social endeavor. Communities of scientists work together on problems defined by their particular fields of inquiry. Each community has a technical language of its own spoken by its adherents. There is usually a general consensus among members of a scientific community regarding the nature of the reality that they are studying and what constitutes an acceptable method of investigating that reality. Scientific communities are often validated and supported by academic institutions, governmental agencies, business interests, news organizations, and popular culture. Contemporary science recognizes these cultural influences and has become quite savvy at using them to its advantage. As a consequence, the lines are so blurred that it is often hard to tell the difference between science, business, politics, and entertainment.

Those who disagree with either consensus reality or proper methodological protocol tend by a variety of means—usually social pressure from both within and outside the scientific community—to be marginalized and made to feel unwelcome. For this reason, credible challenges to scientific consensus are often ignored, not because they are found wanting, but because of scientific groupthink. The same applies to medical science; it is a socially dependent phenomenon that has constructed its own reality to suit its particular needs and, as per the thesis of this book, this reality often comes into direct conflict with the personal beliefs and experiences of the patients that it is intended to serve.

Many scientists believe that because the general public does not have the necessary training or expertise, it therefore does not have the authority to pass judgment on the validity of claims made by scientific communities. While outside forces can provide or withdraw support for scientific endeavors, the professional community itself decides what is acceptable and valid scientific knowledge and what is not. While its ultimate success depends upon cultural acceptance, a scientific community manufactures its own body of scientific knowledge and the same community determines the veracity or falseness of claims made by its members. Biologist, Rupert Sheldrake, explains how the sciences cannot be considered objective simply by virtue of the fact that they are produced by communities of human beings:

> *Sciences are human activities. The assumption that the sciences are uniquely objective not only distorts the public perception of scientists, but affects scientists' perception of themselves. The illusion of objectivity makes scientists prone to deception and self-deception. It works against the noble ideal of seeking truth.*[35]

All of the aforementioned cultural and social dynamics are particularly applicable to modern medicine. They play influential roles in shaping the medical community and its prevailing beliefs. These forces are so prevalent that it is hard to imagine how any truly groundbreaking idea could successfully run the gauntlet of scrutiny to gain acceptance from the medical community. The resulting medical caste system rewards moneyed interests and those who conform to consensus beliefs, and treats independent-minded medical innovators as fringe lunatics—aka charlatans, quacks, and promulgators of "pseudoscience." Such innovators are usually practitioners who also belong to communities of like-minded individuals but who hold to worldviews that differ from that of mainstream medicine. They study, work, and communicate within alternative communities that operate under different sets of rules and assumptions. They often minister to patients who either disagree with conventional medical views regarding health and illness or who have not been satisfied with the results of conventional medical treatment.

Again, it is important to note that not all medical practitioners are closed-minded individuals unwilling to entertain new ideas. A distinction must be made between individual practitioners and medical culture as a whole, which tends to be quite conservative and very slow to change. It is true, however, that while some conventional practitioners may hold unorthodox views, those views tend to be held privately and are not likely to find a receptive ear within mainstream institutions, professional societies, or popular medical journals. Willingly or not, most physicians must abide by consensus beliefs and practices or risk being ostracized by their profession.

Throughout the history of science, many theories that were once thought to provide accurate representations of "reality" because of their success in solving some problem, turned out to be mistaken. They were wrong in the sense that other theories eventually came along to replace them, either because they provided more successful solutions to the problems at hand, or because they more closely approximated reality as defined by science.

Philosophers of science tend to frame the discussion in terms of the practical success of a scientific theory versus its ability to describe or explain reality. It has been argued many times over that the reason why science has been so enthusiastically embraced is because of the practical results that it produces. That cannot be denied. Just one look around at the many amazing scientific and technological wonders of the modern age will quickly confirm this. The same can be said of medical science. Take imaging techniques for instance—they can reveal hidden internal abnormalities such as fractured bones, ruptured joints, clogged arteries, cysts, tumors, and much more. Likewise, the effectiveness of insulin therapy has made life virtually normal for countless individuals with diabetes. And emergency medicine and trauma surgery continue to save lives on a daily basis. There are many wonderful success stories that can bear witness to the practical value of Western medicine.

The question, however, is whether these successes constitute sufficient evidence to validate the conventional medical worldview as defined by its many *–isms*. A cursory glance at the world of physics and its ever-changing theories regarding the universe and how it all works should make clear how science has been much less successful at depicting reality. After all, if science had a lock on reality then it would not keep evolving the way it does. An accurate and thorough understanding of the nature of reality is a much trickier proposition that many would argue is not even possible. And yet Western cultures have pretty much fallen for the illusion that science has brought us closer than ever to a genuine understanding of the ultimate nature of life and the way things really are.

This is plainly evident in the tremendous inflation of the modern scientific ego, which is largely a function of the practical successes that it has undoubtedly achieved. The many successes of science have helped capture the collective imagination of Western culture, and the confidence of the scientific establishment grows in proportion to the faith placed in it by the general populace at large. But, while science may provide a superior understanding of the "material stuff" of life, it has little to offer in terms of the larger questions of existence. Regrettably, as the scientific ego expands it tends to overstep its boundaries. With the expansion of its own sense of self-importance, Manly P. Hall notes, science has outgrown the need to answer to any higher authority:

> *Scientists began to regard themselves as a race apart. By the middle of the nineteenth century, nearly all the departments of science were suffering from an infallibility complex. To the scientist, wise in his own conceits, all that was not science was superstition...Darwin and Huxley were the demigods of the new era, and their solemn pronouncements on everything in general became the gospel of the proletariat. By this time science viewed ancient authority as a poor relation and excommunicated the illustrious ancients from its honor roll. Like a self-made man, science became ashamed of its own origin.*[36]

Science long ago dispensed with philosophy and metaphysics because they were thought to be no longer relevant. Overconfidence in its own beliefs regarding the nature of reality led science to forego the need to reflect upon the validity of its claims and presuppositions. You, too, would cease to examine your own beliefs if you were absolutely certain of your basic assumptions about the way the universe works. This is essentially what science has collectively done. In a bold and conceited feat of circular logic, many scientists believe that the fundamental questions posed long ago by the great thinkers have been settled—settled by science of course! No explanation is offered; it is just assumed to be true.

Contemporary proponents of the superiority of a scientific world-view have recently sought to discredit those who hold certain religious beliefs. Creationism, for example, does not necessarily have to be in conflict with scientific theories of the origins of the universe. Nevertheless, some find it nearly impossible to entertain the notion that the two possibilities can coexist. The dualistic bent of science leads many to believe that it can only be one way or the other. Here, we have an example of science passing judgment on a virtually universal human belief that it has determined by its own scientific authority to be anachronistic, irrational, and "unscientific." Science long ago decided that personal experience and the non-physical dimension of human existence were not worthy of its time and efforts. And yet science essentially maintains a double standard—while it is too proud to stoop to the level of the subjective, thereby refusing to investigate it, science nevertheless reserves the right to dismiss religious and spiritual phenomena as either illusory or at least not relevant to the interests of humankind, nature, or the greater universe. This amounts to an abuse of scientific authority and more of the same circular, self-fulfilling logic.

Returning now to the basic question as to whether science is widely accepted because of its practical successes or because of its grasp on the nature of existence and questions of being, we must conclude that the former is the case. Though the forces of scientific imperialism would have us believe that the latter is also true, this is merely a function of scientific overreach. Science simply cannot answer most of the questions posed by philosophy, religion, and metaphysics. It is in way over its head and needs to return to the solid material ground that it is purportedly designed to study. While it is capable of shedding light on many practical and logistical matters, science as it is currently conceived cannot and will never be able to replace disciplines that are concerned with questions of the nature and purpose of human existence. Furthermore, the very practice of science itself is highly dependent upon how we answer those larger questions of meaning.

Now, let us segue from this general discussion about science by posing the same questions more specifically to medicine. Is Western medicine so highly regarded because of its practical successes or because it is believed to possess the most accurate understanding to date regarding human health and illness? No doubt, most people would answer that its popularity stems from a mixture of both factors; it is successful *and* it understands more about illness than did the medicine of bygone eras.

I, on the other hand, believe this is an overly generous assessment of Western medicine. It is only partially true. It is easy to believe in the successes of medicine when we fail to acknowledge a number of qualifying factors that allow us to arrive at that conclusion. These factors have to do with the various suppositions made by medical science that define the rules according to its particular worldview. Naturally, those living in an age dominated by that worldview are not likely to question its fundamental premises. Nevertheless, those premises define the very differences between reality as conceived by Western medicine and reality according to a more holistic perspective.

When an antibiotic prescribed for a patient with a sinus infection results in complete resolution of symptoms within a matter of days, it is considered a success by conventional standards. Most would agree that the targeted physical condition in this example yielded to the intervention of the physician. We tend to assume that the antibiotic did its job. However, there are plenty of patients with sinus infections whose symptoms resolve in the same short time without taking antibiotics. Assuming that the antibiotic did resolve the sinus infection, what are we to think when the same patient returns to the doctor two months later with another sinus infection? Or, what about the patient who returns to report that she has developed a new tendency toward diarrhea ever since taking the antibiotic that originally helped resolve the sinus infection? What if another patient reports feelings of anxiety, restlessness, and difficulty sleeping ever since completing the same antibiotic for his sinusitis?

From a holistic perspective, these are important questions. After all, how can we consider these successful examples when antibiotics provided only temporary relief and, in some cases, led to more problematic health issues? Considered as a whole, according to the principles of holism, these are examples of palliation at best (one sinus infection resolves, only to be followed by another), and suppression at worst (a more threatening health issue arises after the resolution of a prior, less problematic issue). This is the unavoidable conclusion when we perceive these scenarios from a broader perspective.

Of course, an argument can be made that this is the best we can expect—antibiotics do work but they are not perfect. Since there is no better option, success must be measured in relative terms. From my own medical experience, this is simply false. There are alternative options to treat sinus infections that are not only safer, but can also reduce the likelihood of recurrences. This is an example of how holistic reality can contradict conventional medical reality. When we take into account all possibilities, it is possible that a more effective standard of care for sinus infections can be achieved.

Conventional medical science makes no claim to holism and, in fact, rejects most holistic beliefs and practices as pseudoscientific. The claim to success in the preceding example requires the adoption of a number of rationalizations in order for it to ring true. One would first have to believe that the subsequent sinus infection was a separate condition, having little or nothing to do with the preceding sinus infection. The subsequent onset of diarrhea would also be excluded from the overall assessment of whether the antibiotic prescription was successful or not. And the anxiety would be viewed as a completely different condition, a mental-emotional issue having no connection to the physical sinus infection.

In order to arrive at such conclusions, one must severely limit the acceptable boundaries that define a medical situation. Not surprisingly, medical science is very adept at doing just this. It compartmentalizes issues of health and illness so well that most patients and practitioners fail to make the obvious connections between related medical events. A conventional medical worldview does not account for process. It is a static paradigm that is poorly suited to the ebbing and flowing nature of organic life systems. Subsequent health issues that arise after the resolution of prior ones are simply viewed as new and unrelated issues. Thus, responsibility for the whole is rationalized away in accordance with the parameters defined by materialistic, reductionist, mechanistic medical science.

Taken on its own terms, modern medicine appears to be the most successful model of human health care ever devised. Many believe that it also provides the most accurate picture of medical reality to date. However, I must regrettably argue that, on both counts, Western medicine has failed the test. Medicine can only be deemed successful when viewed with a particular set of blinders that limit one's ability to evaluate its longer-term impact upon the whole person. Its successes can only be justified by focusing attention on the specific problem at hand without considering the larger picture.

Holistically speaking, Western medicine is successful only in a very limited sense. If drug X noticeably reduces the suffering from a migraine it might be considered successful. But when we take into account the larger picture and we find out, for example, that since using drug X the frequency of migraines has actually increased, thereby requiring more frequent dosing with drug X, then we must reconsider our criteria for success. If this were an uncommon occurrence then it would not be an issue, but scenarios of this nature are more the norm than the exception.

This is what differentiates holism from reductionism. Conventional standards of medical success are based on short-term outcomes and targeted treatment of local symptoms. Holism defines success by the impact of an intervention on the whole person—body, heart, mind, and spirit—over the longer term. Medicine is not as successful as it appears to be when judged according to holistic standards. Of course, there is no reason why both standards cannot work together to form a more effective system of healing.

When medicine fails to live up to standards of holistic success, its understanding of reality needs to be reassessed. If much of medicine generates short-term results followed by longer-term complications then it is imperative that it reevaluate its basic assumptions vis-à-vis the conventional –*isms*.

It is reasonable to think that an effective method of healing that generates long-term health and well-being would entail a more accurate conception of medical reality than a system that produces mostly short-term results. This may or may not be true, and it is a question of lesser concern, because the more important issue is how to restore suffering patients to health *and how to keep them healthy*. Nevertheless, I believe that holism entails a more complete understanding of medical reality because, in my experience, it produces healthier, happier patients. It makes intuitive sense that a more accurate understanding of medical reality—the dynamics of how illness takes hold and how healing takes place—would yield a better system of healing.

To be clear, I should note that my use of the word *reality* is intended here in its common, general sense to mean reality as it is seen, felt, heard, *and experienced* by real people in everyday life. I am using it in the same sense as one uses the word *actuality*. I am not using it in its philosophical sense to mean the absolute and universal nature of objective existence as it may theoretically be apart from human thought and consciousness. It amounts to the difference between a diagnostic label in the abstract and the symptomatic reality experienced by a person with that diagnosis. Scientific truth can never be absolute. Absolute truth is a matter better suited to philosophers and metaphysicians.

Medical science produces an abstract and oversimplified interpretation of the information derived from first-hand patient reality. This relatively new conventional medical reality is constructed almost exclusively by the rational mind without conscious assistance from right-brain input. Rather than making use of factors like intuition, heart, creativity, instinct, and meaning, it purposely sets out to exclude these influences from its so-called objective operations.

The conventional medical paradigm of the rational mind is more accurately characterized as an overly analytical dissociated methodology, disconnected from the holistic reality of real people with real problems. It is not possible to ascertain medical truth via the sterile and clinically detached use of logic alone. Holistic medical reality is a complex phenomenon that cannot be approximated using such a limited model of health and healing. Left-brain medical perspectives deny the role of purpose, meaning, mind, and consciousness when it comes to healing human illness. It is not possible to adequately address illness without taking into account a variety of intangible factors.

Medical Wisdom

By now it should be clear that data and information are not equivalent to knowledge, description is not a substitute for explanation, association does not constitute causation, abstract stereotyping produces a weak imitation of first-hand experience, and the speculative application of logic to a problem does not make the conclusions drawn any more legitimate. Yet each one of these misconceptions contributes to the aura of factual certainty and credible authority of the Western medical enterprise.

The one key missing ingredient that medicine seems to be lacking is good old-fashioned common sense. Western science sneers at the very notion, believing common sense to be a quaint and folksy attribute typically associated with the scientifically uneducated person. Something as vague, imprecise, and uncertain as common sense has no place in technical science grounded in rigorous, objective methodology. It is as if science has convinced itself that its self-evident superiority allows it to bypass a key factor that it mistakenly equates with being mediocre and unsophisticated.

Without common sense, the inexorable drive for more data inevitably leads to information overload. Knowledge entails some degree of understanding. It implies familiarity with the details, facts, and issues regarding the subject matter. One can possess a great deal of information regarding a particular topic but that does not imply that one has a working understanding of that topic. Understanding implies that one can make sense out of a group of facts or body of information. To arrive at an understanding is to comprehend the significance of the information at hand, both in terms of the overall picture and the detailed intricacies involved. As French metaphysician, Rene Guenon, notes, knowledge is useless when it is disconnected from all that would lend it a sense of meaning:

> *By seeking to sever the connection of the sciences with any higher principle, under the pretext of assuring their independence, the modern conception robs them of all deeper meaning and even of all real interest from the point of view of knowledge; it can only lead them down a blind alley, by enclosing them, as it does, in a hopelessly limited realm.*[37]

When, in the name of objectivity, science deliberately sets out to reduce information to its supposedly purest and often quantitative form, the resulting data produced is removed from its context and stripped of its relationships to all other information. Without context and relationship, it is not possible to discern meaning, intention, implication, value, or purpose. It makes it impossible to possess genuine understanding. Medical science's dismissive attitude toward common sense places it in an untenable position. Its allegiance to clinically detached objectivity, quantitative outcomes, and value-free rationality predisposes it to underestimate the necessity of good judgment.

Good judgment, discrimination, and street savvy come not from book knowledge, but from the trials and errors that only personal experience can provide. The problem, though, for medical science is that personal experience is a subjective matter, and subjective information is believed to be inferior, unreliable, and even deceptive. The only "real" facts are the objectively true facts of material existence, while the subjective impressions of mind are ephemeral and can never be truly trusted.

This is only true if one subscribes to an outdated paradigm. It is a distortion resulting from the biases of the scientific worldview, biases that have been exaggerated to an unreasonable degree by Western medicine. Medicine makes the error of coveting quantity of information over quality of understanding. This becomes clear when a clinician's experience is challenged by that familiar refrain of skeptics, "Where is your proof? Show me the data." In an odd reversal of standards, the understanding gained through acquired knowledge and first-hand experience is trumped by the abstract quantitative data produced by research studies.

A thorough grasp of the accumulated knowledge of a given field of study should enable one to achieve a solid understanding of the material. There is, however, yet another level of knowledge that entails a great deal more than mere understanding—and that is wisdom. A wise person has the capacity to think and act by making use of a number of faculties other than just rational thought and quantitative analysis. Wisdom involves intangibles like discernment, sound judgment, common sense, and even enlightenment. Such qualities cannot be directly imparted in the same manner that a teacher teaches a lesson. Wisdom is accumulated gradually and painstakingly, through the successes and failures of personal experiences that can span a career or a lifetime.

Wisdom transcends mere scientific understanding because it is broader than that which can be contained within any particular field of specialized knowledge. In addition to the wisdom conveyed by the great works of science, literature, philosophy, and religion, wisdom is often transmitted informally, from person to person, and from generation to generation. Our elders are often our greatest sources of wisdom. Knowledge can be easily passed from one person to another, but the wisdom required to grasp the full significance of that knowledge is either innate or accrued with the passage of time.

Scientists of bygone eras were well-rounded scholarly individuals, schooled in the arts, humanities, and natural sciences. Their modern counterparts are a relatively rare breed. Breadth of knowledge is no longer valued the way it once was. Specialists whose training involves the acquisition of highly detailed in-depth knowledge of narrow areas of study are the most highly regarded professionals of our day. Science behaves in such a way that he wisdom acquired from years of medical experience may be suddenly cast aside in favor of the latest data provided by a new research study. But true wisdom accepts the uncertainty that always comes with health and healing, and is hesitant to the embrace the illusion of certainty that comes with specialized knowledge.

In practical terms, conventional medical information taken at face value, without an appreciation for its meaning and significance, results in poorly conceived strategies of symptom cessation that tend to be suppressive, not curative. Medical wisdom, on the other hand, is capable of grasping the bigger picture, thereby avoiding the pitfalls of superficial approaches to patient care. Medical knowledge without insight and sound judgment leads to unanticipated consequences, while medical wisdom that takes into account local symptoms, personal circumstances, and holistic reality, is a prerequisite for true and lasting healing.

CHAPTER 14

Scientism, Skepticism & Medical Fundamentalism

People who are devout followers of Scientism share a number of quaint dogmas, the most important of which is the one that they don't have any.
  –Isaac Bonewits

All forms of fundamentalism, religious and scientific, regard themselves as self-sufficient, displaying no interest or concern for external challenges to their dogmas. The contamination of science with scientism and of religion with fundamentalism constitutes a lethal infection, which, if left unchecked, is bound to result in the death of its host; and the aftermath of that fatality bears little resemblance to any genuine science or religion.
  –B. Alan Wallace, *The Taboo of Subjectivity*

Science and Scientism

We have the unique experience of living during the most scientifically advanced era in recorded history. The spectacular achievements of science are undeniable. Many are awe-inspiring. Western culture has become so accustomed to its technological innovations that it is almost impossible to conceive of what daily life was like just a hundred years ago. Science is as popular as it is today because of its tremendous successes.

The practical applications of science have given us electrical power, batteries, photography, television, computers, digital technology, skin grafts, artificial joints, insulin in a bottle, space travel, dish washers, refrigerators, and my favorite, the iPod. The list could go on but it is obvious why we place so much trust in the power of science to provide comfort, security, and effective solutions for our basic needs and problems of daily living. We even go so far as to say that science has performed miracles, and we admire and often revere the scientific pioneers who helped engineer those miracles. Scientists and engineers have become indispensible to our 21st century way of living.

The problem, however, is that we sometimes give science more credit than it deserves because we assume that science knows more than it actually does. Many are mesmerized by science's prowess and are inclined, therefore, to believe the inflated predictions of scientists and the promises of science in general. When the value and power of science are unrealistically overestimated, it becomes susceptible to scientism.

Scientism is an exaggerated belief in the knowledge that science provides and the ability to use that knowledge to solve all manner of problems, human and otherwise. Hardcore scientism asserts that scientific knowledge is the only real knowledge; all real knowledge is scientific knowledge. Only science can provide access to truth. According to scientism, all other forms of human inquiry and experience are not to be trusted. Religion, metaphysics, philosophy, ethics, and even psychology are believed to be unreliable, unscientific, and inferior. According to scientism, only science can provide the answers to both practical problems and the larger questions regarding the nature of life and human existence.

Scientism presumes the superiority of reason, objectivity, and logic to be a given. Scientism assumes that rational knowledge is the only valid form of knowledge. Rational knowledge is scientific knowledge. All other forms of knowledge—intuitive, experiential, emotional, spiritual, imaginal—are subjective, untrustworthy, deceptive, and even contemptible.

Scientism asserts that only that which can be studied by science is real. Those who espouse scientistic beliefs are essentially materialists. Only things that can be seen, touched, measured, and quantified by the methods of science are real. All else is imaginary, a deceptive illusion. As would be expected, objective reality is taken to be the only reality. Thought, emotion, and consciousness itself are reduced to a function of the anatomy and chemistry of the brain. Spirit and soul are fanciful creations of scientifically-naïve persons. Scientism takes all of the –*isms* that we have been discussing and enshrines them as not just true but as the only way to arrive at truth.

Most important, scientism is fundamentally imperialistic. Advocates of scientism believe that they have the authority to pass judgment on the entire range of human social, cultural, and political affairs. Scientism is guilty of a double standard. It discounts other fields of knowledge and experience as unworthy of scientific consideration and then proceeds to pass scientific judgment upon them. It assumes the right to speak with confidence and certainty regarding areas of human interest that it deems irrelevant to science.

The origins of scientism date as far back as early seventeenth century Europe and the Scientific Revolution. Philosophical thought was heavily influenced by the many great scientific triumphs of the time. By the nineteenth century, August Comte had founded his philosophy of positivism, which, in turn, borrowed from David Hume's empiricism. The only true knowledge of the world around us, Comte believed, comes from that which can be empirically perceived by the senses. Such philosophies were naturally compatible with the methods of science. They dealt with tangible things of the world that could be perceived and quantified. All other phenomena outside the domain of science and positivism constituted the illusory fabrications of active imaginations.

Positivism held that religious, spiritual, and metaphysical beliefs, ideas, and practices were based in fantasy. In fact, Comte argued that knowledge passes through successive stages of development beginning with the most primitive, theological stage, followed by the metaphysical stage, and ultimately the most advanced scientific or "positive" stage. The clear implication was that all information other than the knowledge derived from observation, experiment, and facts as established by science, is the inferior product of an immature mind. According to positivism, all human knowledge would someday become scientific knowledge.

Scientistic philosophy was given new life in the twentieth century by the influential Vienna Circle, under the new name of *logical positivism*. A group of philosophers convened at the University of Vienna to develop and promote their common belief that the only legitimate knowledge was the scientific knowledge derived from empirical observation and logical analysis. The Circle's goal was the unification of science and the elimination of metaphysics. The pendulum had swung very far indeed since pre-scientific times when the Christian church was the dominant voice in the West on all matters of life, death, and truth.

Up until the 1960's, logical positivism was the principal school of thought in academic philosophy, particularly as it pertained to philosophy of science. Even today, positivism continues to influence the social sciences. Quantitative and statistical methods in sociology, psychology, economics, and political science are the preferred methods of inquiry. This trend is a function of scientistic influence in the sense that it is believed that quantitative methods yield more factual information, which, in turn, gives these fields greater credibility. To be taken seriously in a scientistic age the social sciences downplay qualitative approaches in order to give the impression of being more like hard sciences.

Science was originally conceived as a systematic means to study the phenomena of creation. Eventually, it came to mean the study of the natural world, where natural meant the material world of physical objects. Scientific method's allegiance to objectivity required that it reinvent the very definition of science. Only a quantitative methodology could control and manipulate the objective, tangible, and concrete phenomena of nature. Subjective qualities like color, sound, emotion, meaning, and aesthetics were not compatible with objective science. They could, however, be re-conceptualized as rods and cones, tympanic membranes, and cerebral cortexes. Thus, the material objects associated with intangible phenomena became the preferred topics of scientific investigation.

Nature was thus severed from its connection to the subjective dimension of human experience. This, in spite of the fact that everyday human experience is, first and foremost, qualitative in nature. After all, we perceive colors, not rods and cones. Meaning and value were no longer relevant to scientific inquiry as they had been in the past. Spirit, soul, and consciousness became taboo subjects, relegated to the superstitious speculations of theologians and metaphysicians. Subjective phenomena came to be regarded as that which constituted the world of mere appearances, while the objects studied by science became the objective "reality" of the material world. Science, to this day, continues to masquerade as an impartial means of studying the natural world—impartial, that is, if we can manage to overlook its decidedly scientistic influences.

Virtually any topic can be approached in an organized and systematic manner. We can study art by its historical periods, stylistic phases, thematic content, materials that are used, and so on. But when conventional science elevates objectivity to its present superior status above the subjective, topics like aesthetics, styles, and themes do not qualify as belonging to the scientific domain. By this definition, the study of art is too vague and imprecise to be granted the privileged status of a branch of science.

In order to accommodate its new, more narrow focus, science had to be redefined to mean a method of inquiry that involves experimental testing of hypotheses formulated from observations made of the natural world. The key term here is *natural world*, meaning the world of concrete quantifiable phenomena. This is how most people understand science today.

A great deal of misunderstanding centers on the fact that science and scientism are often conflated. To begin with, most people have never heard of scientism. The distinction between the two is crucial to the theme of this book. I have the utmost respect for science that is practiced with full awareness of its purpose, its limitations, and most importantly, the presuppositions that inform its worldview. But scientific knowledge wielded without perspective, as is the case with scientism, can be likened to using a chainsaw in place of a nail clipper.

Science is supposed to be an *investigative tool* used to discover the intricate details of natural phenomena. Scientism, on the other hand, is an *ideology*. Scientism represents an abuse of scientific authority. Scientism is science without boundaries, science that has gone overboard. The imperialistic impulses of scientism lead it to make judgments that responsible conventional scientists recognize as beyond the scope of their methods of investigation.

When a particular scientist declares, for example, that there is no such thing as a higher power and that all believers are therefore delusional, this is, in fact, a scientistic statement that has no basis in scientific knowledge. While it may be true that it cannot be proven by scientific standards that a god does exist, it also cannot be proven that a god does not exist. Any assertion to the contrary is scientifically groundless. Put another way, any statement that implies that science can determine, either now or someday in the future, that a supreme entity does not exist is a statement that reveals its scientistic bias. It is an expression of belief, not fact.

Advocates of scientism believe that science has already answered or is on the verge of answering the age-old questions of the origins and meaning of human existence. The forgone conclusion is that there is no mysterious meaning, that a higher power or higher form of consciousness does not exist. All that counts and all that exists is the objective reality of the natural world. By definition, nothing exists beyond the objects of science. Such assertions constitute the *a priori* assumptions of the scientistic worldview. Since they cannot be proven, these assumptions are more appropriately characterized as metaphysical beliefs rather than scientific facts, much to the horror and denial of scientistic advocates.

Herein lies perhaps the most critical distinction between science and scientism. Science *chooses* to leave the subjective dimension of human experience out of its observations and experiments, regardless of whether subjective considerations are believed to be of significance or not (and regardless of whether it is even possible to do so). Scientism, on the other hand, *presupposes* without scientific confirmation that the subjective *is* either irrelevant, or does not exist. Scientism has already passed judgment and the conclusion is that the only reality is the reality that can be confirmed by scientific methodology. If science cannot prove it, then it is not true. Scientism, therefore, is the most dogmatic and pernicious of all the medical *–isms*. In this regard, scientism is also guilty of unwavering allegiance to all of the other *–isms* that define conventional science.

The dichotomy between science and scientism is not as black and white as it may seem, because there are varying degrees of scientistic purity. Scientism is not an officially recognized body of academic knowledge, nor is it an avowed philosophy of a particular community of believers. Nevertheless, it is a cultural phenomenon, the exponents of which hold both conscious and unconscious scientistic beliefs that vary from individual to individual. The most ardent followers of scientism believe that science is without boundaries and that someday all problems will be solved by science alone. Furthermore, most advocates of scientism do not see themselves as such. They reject the label of scientism because they simply see themselves as defenders of science. They believe their version of science to be the one true version of science. It sounds a lot like a religious doctrine, does it not?

Problems emerge when the lines between science and scientism become blurred because, for all practical purposes, the modern world tends not to distinguish between the two. Most people cannot tell one from the other and tend to interpret them as if they are both representative of legitimate science. This unfortunately leads to all sorts of misunderstandings. Most of the time, when the general public reacts negatively to the overreach of science, it is reacting to the imperialistic agenda of those who hold scientistic views.

For example, in 2012, when the media trumpeted the news that physicists had finally discovered the "God particle," it did so without realizing that very few serious scientists would concur with such a characterization. It also did so with a lack of sensitivity to those whose religious sensibilities were offended by such a statement. Almost twenty years earlier, Nobel Prize-winning physicist, Leon Lederman, co-authored a book titled *The God Particle: If the Universe Is the Answer, What Is the Question?* In it he writes about the elusive Higgs boson particle, the holy grail of physics:

> *This boson is so central to the state of physics today, so crucial to our final understanding of the structure of matter, yet so elusive, that I have given it a nickname: the God Particle.*[38]

In the same breath, he goes on to compare this quest with a passage from *Genesis* in the Bible that references Babel. In the passage, the Babylonians attempt to build a tower that reaches to heaven. The story's theme can be understood as a human act of defiance in the face of the unknowability of god. Lederman then goes on to state:

> *The issue is whether physicists will be confounded by this puzzle or whether, in contrast to the unhappy Babylonians, we will continue to build the tower and, as Einstein put it, "know the mind of God.*[39]

Regardless of Lederman's original intent, the statement at least implies that physicists may someday discover a particle that will give them god-like knowledge of the workings of the universe. Of course, that particle has now been found and, not surprisingly, physicists are no closer to Lederman's imagined grail. This is a classic example of the hubris of scientistic overreach, which rightly disturbs spiritually-minded individuals who understand that, while science does discover laws that govern physical existence, and is capable of invoking awe at the magnificence of creation, it is not capable of revealing higher spiritual truths.

The same argument applies to the Big Bang theory. If we believe scientistic claims, we must conclude that science will someday discover how the Big Bang occurred, and when that happens we will know how the universe was created. It will embolden scientists to believe that the universe was created through purely material forces that have nothing to do with spiritual forces or any higher power. The catch is that the explanations offered by science always fall short and inevitably lead to new questions, the answers to which remain beyond the reach of the methods of reductionist science. New information just leads to more questions—while the same eternal mysteries persist.

A common rationale used to cover the gaps in scientistic arguments is a belief that can be summarized as, "We haven't figured that out yet but someday we will." Many Westerners embrace this sort of wishful scientistic thinking that promises great rewards somewhere off in the future. Such thinking reinforces the illusion of scientific omnipotence. Such claims have no basis in science—they merely represent the aspirations of scientists. The perpetual promise of a cure for cancer is a perfect example of this carrot and stick mentality that guarantees generous funding for the industry but yields very few genuine therapeutic breakthroughs. Even the use of the word cure in reference to cancer is, in my opinion, an abuse of medical responsibility.

In addition to religious belief and spiritual experience, there are two broad categories of knowledge that lie directly in the crosshairs of scientistic scorn: parapsychology and holistic medicine. Both can legitimately claim to be grounded in scientific method, although not necessarily scientific method as defined by materialism, reductionism, mechanism, and so on. As such, they challenge the cherished beliefs of those who cling to a scientistic worldview. The knowledge provided by paranormal studies and holistic healing poses a direct threat to scientistic dogma.

There is little difference, therefore, between those who represent the scientistic left and some of the more extreme members of the religious right. The further to either pole that one swings, the more dogmatic are the views espoused. Beliefs from both extremes have crept into mainstream channels, largely due to misunderstandings that fail to discern the differences between science and scientism on the one hand, and fundamentalist dogma and spirituality on the other.

It is not unusual for some, upon hearing inflated scientistic claims, to believe that all science is of a similar nature. Hence, the religious right's misguided rejection of an abundance of good science. Likewise, the left dismisses much of the newly emerging spiritual and esoteric knowledge in Western culture because it is mistakenly conflated with religious dogma. Many also naively believe the assertions of mainstream science—which has become increasingly infected by scientistic thinking—when it claims that there is no validity to psychic studies or holistic medicine. Extremist positions on both sides give a false impression as to the true nature of conventional science, on the one hand, and spirituality on the other. This increasingly polarized state of affairs has ignited the modern culture wars between science and religion. In essence, it amounts to two fundamentalist camps laying claim to the alleged one and only truth. Neither faction is willing to admit that there is a great deal of grey area between the two extremes.

This highlights an important point. In the vast unacknowledged gulf between mainstream medical science and orthodox religion lies an enormous wealth of untapped information and wisdom. Beyond the bounds of orthodox and monotheistic religion there are a wide variety of spiritual disciplines, experiences, and phenomena. Many are regarded heretical and even blasphemous. In similar fashion, there are a multitude of viable medical theories, health practices, and therapeutic options that can be found beyond the limits of conventional drug and surgical approaches to illness. These options, too, are considered taboo because they are believed to be unscientific quackery.

At one pole we have orthodox religion, at the other, orthodox medicine. Everything in between constitutes a forbidden zone. Authoritative voices warn that you enter the forbidden zone at your own peril. And yet, personal experience repeatedly confirms that not only are most phenomena in the forbidden zone not dangerous, but they are wondrous opportunities that can lead to spiritual growth, emotional maturity, increased vitality, and greater health. In fact, the midpoint of the gulf between the two polar extremes is where the most balanced and holistic phenomena exist. This is the cutting edge of the new paradigm where black or white dualistic thinking is understood for the folly that it is.

Like a fundamentalist religion, if scientism had its way, it would prohibit us from entering the forbidden zone. Mind and consciousness would be trivialized as by-products of brain anatomy and chemistry. The power of prayer and intention would be laughed out of the clinic. All new and promising therapies would be denied a hearing unless they could pass the partisan litmus test of RCTs and EBM. Patient feedback regarding medical care would be rejected as anecdotal evidence. And medical autocrats who know best would deny patients their rights to personal choice. Although many medical professionals do not fit this mold, the overall trend in medicine as a whole is undeniable. Just ask a handful of patients about their experiences with the medical system. They are not likely to paint a flattering picture.

Skepticism

Contemporary defenders of science as the one true reliable path to knowledge like to call themselves "skeptics." Skepticism is a growing trend made possible by internet access, which has enabled skeptics to communicate more easily amongst themselves. Thus a small number of previously isolated persons have organized into a larger and increasingly vocal faction of malcontents. Skeptics scoff at charges of scientism and genuinely believe that they represent the last bastion of rational purity. They are the defenders of science and reason in an increasingly irrational world. In typical black and white thinking, all who disagree with them are assumed to be either scientifically naïve or just plain anti-science.

Skeptics are the modern-day scientific equivalent of the philosophical positivists of a bygone era. Both scientism and positivism hold forth the proposition that the only truth is truth as it is defined by scientific knowledge. Skeptics genuinely think that their belief system—and I intentionally use the words, "belief system"—is grounded in science, not philosophy. Philosophy is anathema to skeptics. Based on their claims, one would think that their worldview consists solely of settled facts that have been proven by science. There is no wiggle room for philosophical discourse, and certainly not for metaphysical speculation. They simply will not admit that scientific methodology is predicated upon a handful of very important metaphysical presuppositions that can neither be proven nor disproven by science.

Skeptics proudly believe in the primacy of scientific method. Anything that has not been proven by science, or has not been verified by experiment, is assumed to be false until proven otherwise. All knowledge obtained by non-scientific means is inferior or fallacious knowledge. Their world is a very concrete, black and white world. Skeptics are known for their annoyingly persistent demands for proof. By proof, of course, they mean demonstrable scientific evidence that affirms their worldview. In medicine, this means the evidence supplied by randomized controlled trails and evidence-based medicine. Although some skeptics do have scientific backgrounds, many are armchair quarterbacks, untrained in any particular scientific discipline. The bottom line is that they believe in scientific method and its ability to yield the truth, if not now, then eventually. They have supreme confidence that scientific trial and error will eliminate falsehoods, debunk specious claims, and preserve the integrity of the scientific knowledge base.

The modern skeptic movement seems to have taken root in 1976 when academic philosopher Paul Kurtz proposed the establishment of the Committee for the Scientific Investigation of Claims of the Paranormal (CSICOP). In a twist of fate indicative of the high degree of intolerance within skeptic circles, Kurtz was essentially forced out of the movement that he founded in 2010 after he criticized skeptics for their "mean-spirited ridicule"[40] of religion. The term *scientific skepticism* seems to have first originated in the writings of astronomer, Carl Sagan, who argued that...

> ... *science requires the most vigorous and uncompromising skepticism, because the vast majority of ideas are simply wrong, and the only way you can distinguish the right from the wrong, the wheat from the chaff, is by critical experiment and analysis.*[41]

Here Sagan applies the same litmus test to an important tenet of most religious doctrines:

If some good evidence for life after death were announced, I'd be eager to examine it; but it would have to be real scientific data, not mere anecdote … Better the hard truth, I say, than the comforting fantasy.[42]

I suspect that skeptics have taken Sagan's sentiments a bit too literally in service to their cause because his work as a whole does not reflect some of their more extreme beliefs. The evangelical zeal of some skeptics has made them relatively famous, household names. The louder and more presumptuous their claims, the more attention they garner. Noteworthy skeptics include Richard Dawkins, author of *The God Delusion*, James Randi, the magician whose mission is to debunk all forms of New Age belief, and the recently deceased Christopher Hitchens, outspoken journalist and media pundit who was known for his belief that scientific inquiry should be able to serve as a replacement for religion and the role that it plays in society.

Comedian, Bill Maher, is a good example of a popular figure whose moderately scientistic views are quite apparent. Although I tend to sympathize with his liberal leanings, I am often struck by his intolerance toward anything that resembles a form of religious belief. At the same time, he exhibits near reverence for the great American myth of scientific progress. His closed-minded attitude toward metaphysical propositions of any sort causes him to conflate all spiritual beliefs with religious dogma. The telltale sign of his scientistic beliefs is that he is as dogmatic in his scientistic views as the religious dogmas that he criticizes.

As in Maher's case, there seems to be a strong correlation between skepticism and atheism. Many skeptics are militant atheists who believe it is their duty to rid the world of its religious fantasies. Interestingly and to his credit, Maher is not a critic of holistic medicine. In fact, he regularly critiques the shortcomings and dangers of conventional medicine.

Another popular cultural torchbearer for the scientific cause who makes frequent television appearances is astrophysicist Neil deGrasse Tyson, Director of the Hayden Planetarium. His scientistic leaning is becoming increasingly apparent. His public persona is aimed more at the perfectly reasonable goal of promoting scientific literacy rather than debunking non-scientific perspectives. But every once in a while he can let slip a doozy. The following quote illustrates the extent to which he is willing to discredit direct experience in favor of conventional scientific standards of evidence:

> As a scientist, I need something better than your eye-witness testimony. Because if even in the court of law eye-witness testimony is a high form of evidence, in the court of science it is the lowest form of evidence you could possibly put forth.[43]

Skepticism has become an organized force with growing influence, especially in the United Kingdom and Australia. Skeptics there have waged active campaigns against complementary and alternative medicine, which they claim has no science to support it. The paradoxical clue that exposes the conspicuously unscientific mindset of skeptics is the extent to which they will go when denying the evidence placed before them. In spite of mounting research as to the benefits of a variety of holistic modalities, skeptics persist in their claims to the contrary. Like ostriches with their heads in the sand, they will do anything in their power to ignore research studies that contradict their beliefs. When it can no longer be ignored, skeptics relentlessly find fault with the evidence in the nitpickiest of ways. I would not begrudge them their cynicism if skeptics applied the same degree of scrutiny to conventional medical practices but, alas, they do not. And it is disturbing when their assertions infect the views of actual scientists and the general public, both of whom may not consciously discern the differences between scientific and scientistic claims.

A prime example of this disruptive influence resulted in a firestorm of protest in the spring of 2013 after TED Conference organizers bowed to pressure from skeptics. As a consequence, TED's website removed video recordings of two prominent speakers at a TEDx event in London billed as "Challenging Existing Paradigms." Historically, TED conferences have provided an influential forum for cutting edge ideas that may not otherwise receive a fair hearing in the public arena. TEDx refers to international events that are sanctioned by the TED organization. The acronym stands for "Technology, Entertainment, and Design" and TED's tagline, "ideas worth spreading," proves to be more than a bit ironic in the context of this dust-up.

Celebrated English biologist, Rupert Sheldrake, whose talk, in accordance with the stated purpose of the event, was intentionally designed to challenge scientific materialism, wound up being accused by TED of promulgating pseudoscientific ideas. The organization subsequently sent a memo[44] to the TEDx community of organizers regarding the need to be on the lookout for would-be promoters of "bad science." TED's knee-jerk response to the potential diminution of its reputation was to circle the wagons against all forms of "pseudoscience" and "health hoaxes."

The TED memo spelled out the "marks of good science," which included the criterion that "it does not fly in the face of the broad existing body of scientific knowledge." I do recall that quantum physics was once in the very same position but, now, is so mainstream that to deny its place would be to commit heresy. The memo went on to define the "marks of bad science," noting that it "has failed to convince many mainstream scientists," "comes from overconfident fringe experts," uses "imprecise, spiritual or new age vocabulary, to form new, completely untested theories," and the clincher, "speaks dismissively of mainstream science." There seemed to be an awful lot of emphasis on defending the status quo and dismissing ideas that do not conform to the prevailing standards of conventional medical science.

The memo then listed the warning signs of a speaker who is not up to TEDx standards, including if a speaker's "affiliated university does not have a solid reputation." Reputation for what, mainstream thinking? Sheldrake has a long list of credentials that includes both Harvard and Cambridge. Speakers were also to be considered suspect if "there is little or no comment on them in mainstream science publications or even on Wikipedia." Now, it is well known in holistic medical circles that the people behind Wikipedia include some avowed skeptics with a rabid anti-alternative medicine bias. The final irony was the TED warning that it is a "BIG RED FLAG" when speakers sell products or services related to their talks. Please forgive my sarcasm, but I don't suppose they were referring to PhRMA here, were they?

In closing, the memo stated, "Bad science talks affect the credibility of TED and TEDx: it is important we get this right." In a single fit of hysteria over its precious reputation, TED had done an enormous disservice to countless individuals whose careers are dedicated to the cutting edge of the newly emerging scientific paradigm. In the process, TED had gone running straight into the arms of the left wing fringe of medical scientism. The skeptics had achieved their goal of stifling the free and open discussion of new ideas in the scientific arena—and they did so by accusing those ideas of being unscientific.

In the end, sanity prevailed and TED was forced to retract its actions, although it did so half-heartedly after the damage had already been done. There is no doubt that organizers were forced to think twice about who they would invite to speak at future TEDx conferences. On a positive note, this skirmish helped bring to light the toxic impact of skeptic organizations on open debate regarding important cultural issues.

I have personally encountered skeptics in the comment sections of my *Huffington Post* articles. They seem not to have noticed the first few articles but, once they caught on to the fact that my topic of choice was holistic medicine, they swarmed like piranha. They attacked each new article regardless of its content, as if determined to intimidate me into ceasing my heretical claims. They struck me more as hecklers and stalkers rather than the defenders of reason that they purport themselves to be. When presented with a variety of forms of evidence, they predictably found fault. They are experts in the minutiae of RCT protocols and will reject anything that does not meet their imaginary standards of conventional medical perfection.

I learned quickly that it is not possible to have a fair and reasonable debate with skeptics. Their sole objective is to win an argument. Their views are preset in stone. They already have all the answers and are not interested in what someone with expertise beyond their level of understanding has to say. They are adept at sophistry and notorious for their circular argumentation. To engage them in debate is a futile exercise. When challenged, they claim the mantle of scientific fact to defend their views—as if merely invoking the word "science" renders them immune from fallibility. When backed into a corner, some actually have the gall to shamelessly claim that the assertions of holistic thinking are false simply because they are "not plausible." In that moment, they abandon science and resort to an argument that essentially says, "if it doesn't make conventional scientific sense to me, then it just can't be possible, period, end of discussion."

Of course, the history of science is riddled with theories that were believed to be unequivocal and definitive, only to be later revised or overturned as scientific understanding evolved through changing paradigms. Skeptics seem to think that history does not apply to them, that science is a settled matter, and that it is not possible to learn something new that might cause them to reconsider their basic beliefs. Their goal is not to inquire like good scientists do, but to discredit anything that threatens their imaginary supremacy. Without knowing it, they themselves are anti-science. They are ideologues disguised as scientists.

Having experienced opposition from skeptics firsthand, I am familiar with a number of tricks that they employ when trying to win an argument. Their lack of knowledge about my unusual medical specialty is clearly evident, but that does not stop them from making a variety of false assumptions about my profession. When I point out their errors it does not prompt them to ask questions that would enable them to rectify their misconceptions.

Predictably, skeptics' first tactic is to demand to see research studies that back up any claims being made, knowing full well that in most new fields of inquiry, as is the case with many holistic modalities, one would expect to find very few, if any studies. When the research demanded is not produced, skeptics are quick to declare victory, as if this constitutes evidence in support of their argument. On the other hand, it does not seem to occur to them that some holistic methods have been around for centuries and their credibility is based not on modern RCTs, but on a long history of empirical evidence and actual case studies. Regardless, this type of evidence does not matter to them because their intent is to invalidate by any means possible.

When skeptics' attention is directed to the fact that there *are* an increasing number of well-conducted studies that *do* support the clinical results of my own homeopathic modality, they immediately dismiss those studies as flawed and inferior. All sorts of disingenuous arguments are made to deny the evidence, including claims that a study is flawed, is not placebo-controlled, involved too few subjects, or the results were not significant enough. When they are made aware of a study that supports a particular alternative modality, they are quick to reference another study that disputes those findings. The situation quickly degenerates into a game of "my studies are better than yours." When the evidence supporting a holistic therapy is particularly compelling, the claim that "more studies are needed" is the final gambit used in the hopes of shutting down the debate. I have no doubt that if skeptics were to apply the same critical eye to conventional medical research, the entire foundation of EBM would crumble to dust.

Many defenders of holism are seduced into playing this unwinnable game. They are tricked into believing that if they can just amass enough evidence then they will eventually be taken seriously. They fail to understand that the game is rigged, that holistic principles cannot be validated through studies designed to meet reductionistic, mechanistic, and materialistic standards. Even when solid evidence that meets conventional medical standards is produced, it is not likely to be believed. Sadly, the net effect is that much research into alternative therapies is geared toward convincing skeptics rather than adding to the body of knowledge of the particular modality.

When skeptics feel threatened they can resort to invective. Name-calling is always a sure sign that someone is on the defensive. In times past, physicians who thought outside the box risked being called charlatans and purveyors of snake oil. Sometimes it was true, but there are many valid alternative therapies, once the object of derision, that are now virtually mainstream. Osteopathy, chiropractic, and acupuncture are a few such therapies. Nowadays, attacks directed at innovative ideas are more sophisticated. *Pseudoscience* is the modern day scientific-sounding euphemism for quackery. The word has found its way into the popular lexicon and it implies that something is not supported by conventional science. When the general public hears that something has been labeled pseudoscience, it assumes, often wrongly, that it has been tested and found to be ineffective. This is simply not the case. The term, pseudoscience, is the medical equivalent of a bigoted slur, and such slurs are almost always grounded in fear and willful ignorance.

In actuality, pseudoscience does not exist—other than in the paranoid minds of skeptics who believe that their cherished idea of science is under assault from imposters. They fail to understand that there are a variety of unorthodox scientific disciplines that represent different approaches to scientific method. There are hard sciences like chemistry, geology, and oceanography, and softer sciences like psychology, anthropology, and sociology. As far as I am concerned, there are also more speculative sciences like astrology, which studies the relationships between the movements of the heavens and human behavior, and alchemy, which studies the relationships between mind and matter. They are all genuine fields of inquiry that sometimes use different methods of systematic discovery. In the same vein, anatomy and pathology are hard medical sciences, psychiatry and physical therapy are softer medical sciences, and music and art therapy are more speculative medical sciences.

Harder sciences are more physical, reproducible, and amenable to mathematical modeling than are soft sciences. This does not make them any more "scientific" than soft sciences. It also does not make them any more believable, certain, or true. Herein lies the folly of skeptical criticism. A harder science may be more certain as to the materiality of its subject matter but this is not a criterion that defines science as a whole. It simply represents the material bias of conventional science. Skeptics take that material bias to an extreme and claim that the physical world is the only knowable world. Some skeptics would deny this bias, but that would be incongruent with the arguments that they make in their zeal to invalidate paranormal phenomena and all forms of alternative medicine. Pseudoscience, therefore, is nothing more than a slur dreamed up by those who believe in an overly rigid definition of science.

Junk science is another similar slur that has made its way into the mainstream. Like pseudoscience, it is used irresponsibly by those with an ideological axe to grind. Perhaps the most insidious medical slur of all is the term *anecdote*. Skeptics use the word like a mantra, perhaps with the belief that if they say it enough it will turn out to be true. The word, anecdote, may sound technical and sophisticated but its sole purpose is to give the impression that direct first-hand experience does not qualify as sound evidence. If we took this idea seriously, then how could we trust the words of anyone, including doctors, patients, scientists, and researchers? A doctor would not be able to trust his or her patients, and physicians would be unable to trust the clinical experiences of other physicians. And skeptics would agree, claiming that this is why randomized controlled trials are superior. Personal experience is believed to be anecdotal because it is subjective, while quantitative data is superior because it is objective, thus making it compatible with scientific method.

To skeptics, the numbers are what count, regardless of the fact, for example, that a person whose lipid profile improves through the use of statin drug therapy can still have a heart attack or develop liver toxicity from that drug. The numbers rarely give a full picture of patient reality. For the very same reason that RCTs are trusted, I find them to be biased. They selectively take into account a few parameters while ignoring all others. The so-called data derives from a number of sources including lab values, diagnostic findings, diagnostic imaging, physician observations, and patient reporting. All of these factors, including interpretations of lab values and diagnostic images have subjective components to greater or lesser extent. Thus, by definition, the data includes a great deal of anecdotal information. Undaunted, skeptics deny this reality and argue that data has nothing to do with anecdote.

There is a fascinating bit of history behind this idea that anecdotal information does not constitute reliable evidence. Raymond Wolfinger, PhD, Professor Emeritus at the University of California at Berkeley, once confirmed a quote[45] attributed to himself. He recalled that while teaching a graduate seminar in the late nineteen sixties at Stanford University, one student dismissed another student's statement as "mere anecdote." The professor responded at the time by stating, "the plural of anecdote is data," meaning that the accumulation of a number of personal observations does rightly constitute a collection of data.

This is particularly relevant to our discussion because skeptics have been known to reflexively rebuke arguments supporting the validity of personal experience with the charge that "the plural of anecdote is *not* data." As far as I can figure, what skeptics really mean to say is that the conclusions drawn from logical analysis are superior to direct experience. It is a product of their left-brain mode of perception and their belief that the rational function provides a more accurate representation of reality than any right-brain organic perspective.

An anecdote was once understood to be a short story or brief description of personal experience. Nowadays, the scientific importance of personal experience is minimized, and a great number of people have been fooled by the scientistic claim that anecdotal information cannot be trusted. By this definition, even a physician's story about a patient's response to a particular therapeutic intervention is an anecdote that is to be regarded with suspicion. Skeptics like to paint a picture of anecdotal evidence as dangerous pseudoscience that will lead science down the path of chaos and dissolution. If we believe anecdotes, they say, then it is just a slippery slope until we believe any claim whatsoever. Now that might be true in the unlikely event that we were to lose track of our common sense, good judgment, and capacity for critical thinking. But such a scenario has little bearing on reality, except in the case of scientism, because this is precisely what skeptics do when they insist that their version of scientific method is the only reliable path to knowledge.

Anecdote is now understood by most health care professionals to mean unreliable hearsay. Repetition of this misnomer has also had a brainwashing effect on the lay public. Patients have been conditioned to distrust their personal assessments as to their own states of health. This, for example, is what causes many to rationalize away or tolerate the disturbing side effects of some drugs, in spite of the first-hand warning signs that they may be experiencing.

The notion that anecdotal information poses a threat to scientific rigor is just plain nonsense. If applied evenly to all parties involved, then no statement by any patient or physician could ever be taken seriously. The use of the term, anecdote, in medical settings is pejorative and disrespectful to patients whose first-hand reporting of their own symptoms and responses to therapeutic interventions is a valuable source of information.

Physicians, too, have been conned into believing that test results and research findings are superior to the lessons gleaned from their interactions with patients. Nothing could be further from the truth. No amount of book learning or research data can replace the clinical wisdom that can only be obtained through the trial and error of first-hand experiences between doctor and patient. The philosophical miscalculation that has caused science to trivialize personal experience is part of the imbalance that contributes to the sickness of modern medicine. I believe that it is time for physicians to reclaim their personal authority and autonomy by rejecting the worst scientistic impulses of the profession that would denigrate their clinical experiences as merely anecdotal.

Another term that has become vulnerable to scientistic abuse is *inconclusive evidence*, especially when it is used to downplay the significance of some finding that threatens the scientific status quo. There is no doubt that many situations in both clinical and research settings legitimately call for further information before a well-reasoned conclusion can be made. On the other hand, it is common for defenders of scientistic dogma to argue that the research findings of holistic therapies constitute insufficient evidence and are therefore inconclusive. In such cases, the call for more studies is, in reality, a disingenuous ploy intended to discredit the evidence at hand.

One would think that reports of therapeutic successes would rouse the curiosity of open-minded health care providers, rather than elicit boilerplate responses designed to dampen the enthusiasm that should come with the possibility of discovering new ways of helping patients. Make no mistake, just like *anecdotal, inconclusive evidence* has become a virtual mantra in scientistic circles. It is used primarily to refute ideas and practices that challenge conventional medical understanding. The term should be viewed with suspicion wherever it arises.

Another turn of phrase that has been stretched far beyond its original meaning is the *risks versus the benefits*. This expression is subject to abuse more so in mainstream scientific usage than in scientistic circles. Originally used as an expression of the pros and cons of a particular therapy or intervention, it is now commonly used to justify the benefits *regardless of the risks*. It turns the physician's calling to do no harm on its head. When we are told that the benefits outweigh the risks, it is important that this not be taken at face value. Further investigation often reveals that those risks are more serious than one would think.

It begs the question as to who should be the judge of acceptable risk. In my estimation, it should be patients who get to decide—provided that they are given honest and accurate information regarding those risks. Unfortunately, such information is hard to come by given the built-in bias in favor of conventional medical therapies and deceptive assessments in terms of their risks and benefits. My years of practice in holistic medicine have given me a unique perspective regarding relative risks. It is a perspective not accessible to most conventional physicians who are familiar only with mainstream options. From my vantage point, the risks of many conventional therapies are magnified when compared to a variety of safer, and sometimes just as effective, alternative modalities. This comparative perspective significantly changes the calculus when weighing the risks versus the benefits.

"Risks versus benefits" has come to mean something very different than was originally intended. We might just as well say that all conventional therapies are safe until proven otherwise. PhRMA is understandably pleased with this status quo because it implies that their products are innocent until proven guilty. The use of terms like *anecdotal, inconclu-*

sive evidence, and *risks versus benefits* has more to do with medical politics than with medical science.

When all else fails and skeptics are unable to refute the evidence in favor of some alternative modality, they have been known to resort to the Hail Mary pass. This is where they insist that it is just not possible for a therapy to work because it contradicts the laws of nature. Since the mechanism of action of the therapy in question cannot be explained in conventional scientific terms, it is assumed that it must defy the basic laws of biology, chemistry, and/or physics, thereby rendering it factually impossible.

Dr. Peter Fisher and his cohorts at the Royal London Hospital for Integrated Medicine have coined the phrase *plausibility bias*[46] to refer to this tendency for skeptics to allow their scientistic allegiance to conventional medicine to color their vision. Homeopathic medicine provides the prime example here where, in spite of favorable evidence in the lab, in the clinic, and in academic research, the inability to explain how or why homeopathy works infuriates skeptics, causing them to argue that the results are just "not plausible."

Homeopaths argue back that the clinical results are self-evident to any open-minded observer, regardless of whether its mechanism of action can be explained. When they offer provisional explanations for homeopathy based in the language of energetics and physics, skeptics cry foul. In essence, they demand to understand the phenomenon on their own terms before they are willing to acknowledge its existence. When reminded that a good number of conventional drugs have no known mechanism of action, skeptics remain undaunted, unable to recognize their own hypocrisy. Round and round the argument goes, with skeptics determined to discredit homeopathy simply because it is not compatible with their dogmatic scientistic worldview. If it does not make sense, then it is not plausible, and should be rejected in spite of thousands of claims of effectiveness to the contrary.

Medical Fundamentalism

The hallmark of Western science is that it prides itself on knowing things with a high degree of certainty. The general public tends to pay little heed to how science knows what it knows. Most people simply accept science's claims to knowledge because, after all, it is science! Similarly, we tend not to question the findings of medicine because we assume that its assertions are true.

Westerners, and left-brain dominant individuals in particular, seem to have a propensity for thinking in terms of absolutes. Not only does science need to know things but it needs to know them with certainty. The certainty of science is often lauded over other less concrete fields of knowledge as evidence of its superiority. The less certain a given field of endeavor, the less science wants to have anything to do with it. The esteem with which poets, history teachers, and cardiac surgeons are held by society is at least partially attributable to the degree of certainty represented by the nature of the work that they do.

Since the time of Plato, philosophers have been preoccupied with the idea that certain knowledge is the only true form of knowledge. Philosophy of science and epistemology, the branch of philosophy that concerns itself with how we come to know things, have been fixated on the notion that the only true knowledge is knowledge that is *universal* and *certain*. For *Plato, universal* knowledge is unchanging *knowledge* that is and always has been true. Universal and certain knowledge, therefore, is a pretty high standard to strive for—one could say, an almost god-like standard. The fact that modern science and medicine take their cue from this lofty philosophical aspiration has created a great deal of misunderstanding and conflict.

The well-established authority of science came under fire in the 1960's, especially in academia. It was questioned whether objectivity was even possible, and critics suggested that objectivity was actually a value held by a biased scientific elite. The idea that science could be a neutral activity having little to do with human values was rejected. The idea that science was capable of producing universal and certain knowledge, which was presumably superior to all other forms of knowledge, caused an intellectual revolt that manifested as the "science wars" within educational institutions.

Claims to the absolute certainty of scientific knowledge can be understood as the impetus that encouraged the spread of scientistic thinking. I believe that the backlash against science represented by the science wars can be seen as a reasonable protest against scientistic overreach and its presumed monopoly on truth. So as not to be misunderstood, allow me to reiterate that science practiced with awareness of its limitations is not the target of my discussion here. Unfortunately, a great deal of modern science is tainted by conscious and unconscious scientistic dogma and, so, the boundaries are often blurred. Although science and scientism are two separate things in theory, in practical actuality there is much overlap, especially in medicine, and so they are not as easily separated as one would think.

An opposing philosophical view holds that all knowledge is contingent upon circumstance and therefore not universal. It is probable at best and therefore uncertain. Of course, scientific reasoning is not possible without a priori assumptions and the various hypotheses, axioms, principles, and theories that it depends upon. The left-brain mindset recoils at this notion, refusing to concede that scientific knowledge could be so ambiguous and open to question. Remember, however, that Western science is a product of dualistic thinking. In order to make science work, its proponents had to eliminate the prime sources of ambiguity. All subjective phenomena including mind, spirit, metaphysics, moral considerations, and consciousness were determined beforehand, according to the rules of the game, to be off limits.

I believe that this schizophrenic split within science itself is the true source of the science wars. Science claims to be able to produce certain knowledge but, at the same time, is wholly dependent upon the exclusion of all human factors that would taint it with uncertainty. Science has become a self-fulfilling methodology that arranges the rules in such a way as to produce the facts that it desires. The war between science as the defender of truth and those who question its motives and certitude springs directly from this dualistic ambiguity of science itself. Because the left-brain mode is so prone to black and white thinking, it perceives the questioning of its certainty as an assault upon the honor of all science. A small acknowledgment of the presuppositions that inform the scientific worldview would go a long way toward resolving the culture wars that they engender.

This need for certainty is one of the primary motivating factors behind the scientistic thought process. Scientism assumes a world of hard dependable fact grounded in materialistic assumptions. According to scientism, only the most advanced thinkers understand this. They are the ones who see through the immature and superstitious ideas promulgated by religion, theology, spirituality, parapsychology, holistic medicine, and pseudoscience. Scientistic skepticism takes this absolutism to such an extreme that, in many ways, it resembles the most dogmatic of fundamentalist faiths.

At bottom, the true psychological dynamic that underlies almost all fundamentalist assertions is fear. It is the main reason why these defenders of science dig their heels in so hard. In direct contradiction to their claimed allegiance to reason, skeptics' fear of ambiguity causes them to stake out highly irrational positions that have little relation to science or reality. Skeptics are known for their absolute conviction as to the fallaciousness of all forms of religious belief. And, yet, science is incapable of providing even a shred of evidence for or against the existence of a god. Skeptics are equally certain as to the non-existence of paranormal phenomena and psychic abilities. Again, conventional science has little to say on the matter. And in spite of voluminous evidence to the contrary, skeptics refuse to acknowledge the many benefits attributed to a wide range of alternative healing modalities.

Fear of the unknown and the fear of being wrong lie at the root of scientistic thinking. Any admission of doubt could lead to the collapse of the scientistic worldview, which assumes skeptics are on the side of truth and that life is just as black and white as their conception of science. New ideas are taboo because science has already settled all the big issues. The psychological need for certainty prohibits any possibility of open-minded inquiry. This is why scientistic thinking, to the extent that it makes its way into the mainstream, poses such a serious threat to genuine science.

There are those who believe that skeptics' inability to admit doubt is related to their fear of the feminine. This is consistent with my own observations in *Green Medicine* where I explain that Western science operates from an almost exclusively masculine, left-brain perspective. Creativity, emotion, intuition, imagination, the ability to perceive connections and patterns, and the ability to feel at home with ambiguity and uncertainty are predominantly feminine, right-brain characteristics. The right-brain qualities embodied by holistic principles are unfamiliar territory to many scientists and physicians.

Although the demographics are changing, it is no coincidence that conventional science has historically been a male-dominated activity. It is understandable then that skeptics' organizations populated by the most fanatical defenders of science are comprised almost exclusively of men. I must remind the reader, here, that scientism is not gender specific. Both men and women are capable of operating in right and/or left-brain modes. Scientistic fear of the feminine is not to be equated with fear of women—it is a fear of embodying feminine qualities as defined above. Scientism is a masculine phenomenon that is heavily influenced by patriarchal bias.

Not surprisingly, all forms of philosophical discourse are anathema to skeptics. Freethinking is frowned upon in the fact-based world of scientistic certainty. Philosophical inquiry can easily rattle the psychological security that comes with knowing that one is in the right. Skeptics rarely acknowledge philosophical points made in argument and will always seek to turn the discussion back to the facts as they see them. As far as skeptics are concerned, their worldview has been the consensus worldview for decades. They do not seem to understand that the crux of the modern dilemma that divides mainstream medicine from millions of grassroots supporters of holism is a profound difference in worldviews regarding what constitutes health, healing, and medicine.

In his online blog, Stephen Bond, a disavowed skeptic who nevertheless still believes in the superiority of a scientific worldview, explains why skeptics refuse to acknowledge the positivist philosophy that is most closely associated with their scientistic beliefs:

> *One reason you don't hear about positivism often in skeptic circles is that skeptics have no time for philosophy; many skeptics hate and fear it. It's the skeptic Kryptonite. As a fundamental, rigorous, intellectually respectable but defiantly non-scientific discipline, philosophy makes a lot of skeptics feel threatened. Skeptics are like a naval fortress, with weapons fixed to sea; while they regard themselves invulnerable against fleets of art grads, paranormalists, and true believers, they know that philosophers can strike them freely in their defenseless rear. Little wonder that philosophers bring out their inferiority complex. Some skeptics would love to dismiss philosophy, all philosophy, in the same way they dismiss religion, but they'd be afraid of appearing stupid or attracting ridicule in doing so. If anything, they're afraid philosophers already find them ridiculous.*[47]

The same applies to parapsychology and occult phenomena—only more so. Spirituality, mysticism, psychic studies, and supernatural phenomena rouse both fear and hatred among skeptics. They are terrified of the paranormal. It is well known that fear often generates angry and defensive responses. There is little difference between the hatred that fundamentalist religions have for heretics and idolaters and the scorn that skeptics heap upon those who challenge their unqualified belief in materialistic science.

In an article published by the *Journal of Scientific Exploration*, a periodical dedicated to the investigation of topics ignored by mainstream science, L. David Leiter writes about his impressions after having infiltrated a skeptic organization called PhACT (Philadelphia Association for Critical Thinking) and befriending some of its members:

> *The theme that has emerged time after time, as I become closely acquainted with individual PhACT members is this: Each one who*

has disclosed personal details of their formative years, say up until their early 20's, has had an unfortunate experience with a faith-based philosophy, most often a conventional major religion. ... Very often, their family or community has (almost forcibly) imposed this philosophy on them from a very early age; but then as they matured, they threw off this philosophy with a vengeance, vowing at a soul level never to be so victimized again. ... Thus, they gravitate to what appears to them to be the ultimate non-faith-based philosophy, Science. Unfortunately, while they loudly proclaim their righteousness, based on their professed adherence to "hard science," they do so with the one thing no true scientist can afford to possess, a closed mind. Instead of becoming scientifically minded, they become adherents of scientism, the belief system in which science and only science has all the answers to everything.[48]

Any reasonable person should be able to hold two seemingly contradictory ideas in mind at the same time. Many practitioners of science do just that. They have no problem practicing their profession while simultaneously holding personal religious or spiritual beliefs that may conflict with views held by some of their colleagues. Some find no conflict between science and their spiritual beliefs at all, while others compartmentalize the two in order to manage the cognitive dissonance that can result from their intermingling.

Skeptics, on the other hand, have no tolerance for spirituality, primarily because their ideology forbids it. Most skeptics are either atheists or agnostics. Skepticism is defined by its materialist agenda, and atheism is congruent with materialism, which posits that material reality is the only reality. I have no doubt that skeptics gravitate toward their scientistic beliefs because they provide the same sort of psychological comfort and security that religion and spirituality can provide for others. Scientistic belief in atheistic materialism is no less faith-based than many religious doctrines. The world becomes a less frightening place when all things have concrete logical explanations.

Scientism masquerading as science is the newest religion vying for attention in Western societies—and skeptics who promote their scientistic worldviews are the main advocates for the new religion. Their general closed-mindedness, intolerance, and inflexible beliefs rank them right up there with the most dogmatic fundamentalist religions. Skeptical scientism is a mirror image complement to absolutist dogmas that claim to possess the one true key to salvation. They represent two sides of the same uncompromising coin. Scientism's materialist certainty as to the absolute non-existence of a god, psyche, soul, consciousness, or an afterlife forms a direct parallel to fundamentalist certitude to the contrary. It is also a contributing factor to that pernicious strain of consumer culture that finds little purpose or meaning in anything other than the accumulation of material goods.

Skeptic's militant anti-religiosity rises to the level of a faith-based belief system itself and its influence on medicine, and culture in general, should not be underestimated. There are some who have proudly called this growing trend the "New Atheism." *Wikipedia*, the popular online encyclopedia, is known by exponents of new paradigm thinking for its distinctly scientistic bent. Here is an excerpt from *Wikipedia's* entry on the New Atheism:

> *New Atheism is the name given to the ideas promoted by a collection of modern atheist writers who have advocated the view that "religion should not simply be tolerated but should be countered, criticized, and exposed by rational argument wherever its influence arises."… The New Atheists write mainly from a scientific perspective. Unlike previous writers, many of whom thought that science was indifferent, or even incapable of dealing with the "God" concept, (Richard) Dawkins argues to the contrary, claiming the "God Hypothesis" is a valid scientific hypothesis, having effects in the physical universe, and like any other hypothesis can be tested and falsified. … Both Dawkins and (Victor) Stenger… argue that naturalism is sufficient to explain everything we observe in the universe, from the most distant galaxies to the origin of life, species, and even the inner workings of the brain and consciousness. Nowhere, they*

261

argue, is it necessary to introduce God or the supernatural to understand reality.[49]

With evangelical zeal, scientistic proponents of the New Atheism are committed to the eventual eradication of all religious thought, which they believe will someday be replaced by science and the voice of reason. Here, in response to a *Newsweek* article examining the New Atheism, skeptic blogger, PZ Myers, defends scientistic atheism against charges of fundamentalist leanings:

> *The "new atheism"... is about taking a core set of principles that have proven themselves powerful and useful in the scientific world—you've probably noticed that many of these uppity atheists are coming out of a scientific background—and insisting that they also apply to everything else people do. These principles are a reliance on natural causes and demanding explanations in terms of the real world, with a documentary chain of evidence, that anyone can examine. The virtues are critical thinking, flexibility, openness, verification, and evidence. The sins are dogma, faith, tradition, revelation, superstition, and the supernatural. There is no holy writ, and a central idea is that everything must be open to rational, evidence-based criticism—it's the opposite of fundamentalism.*[50]

Here we have an example of scientistic double-talk, wherein the pledge to open-mindedness is accompanied by a qualification of the terms that define the limits of inquiry. Among others, those terms include the necessity of natural causation, a demand for "real world" concrete explanations, the exclusion of personal revelation as a form of knowledge, and that evidence be available for examination by anyone, which is to say that subjective information and first-hand experience do not qualify as evidence. In other words, only that which can be proven by scientism's definition of science is real.

Of course, skeptics' belief that their knowledge is grounded in sound science is a delusion. It is more fitting to say that scientistic knowledge derives from an overreliance on logic to the exclusion of common sense and good judgment. Skeptics equate the left-brain mode of rational thinking and logical analysis with science itself. Right-brain functions have no place within this narrow definition of scientific discovery. Science is associated with logic—no matter how ungrounded in reality and experience the argument may be. In this sense, scientism represents rationality taken to an irrational extreme.

There is also a double standard regarding the acceptable limits of logic and reasoned analysis, which reveals skepticism's outright bigotry. The multitude of ever-changing speculative explanations for the phenomena encountered by theoretical physicists, for example, are fair game to skeptics, while similar attempts to account for the phenomena witnessed by practitioners of holistic therapies are labeled pseudoscience. The energy meridians of acupuncture, the bioenergetic life force proposed by homeopaths, and Rupert Sheldrake's theory of morphic resonance are hypocritically dismissed out of hand because, for some unexplained reason, those hypotheses are not believed to be scientific, regardless of the fact that they are formulated from the clinical observations and experiences of scientists and medical professionals.

It has always been my understanding that true science remains open to all possibilities. It is not characterized by a cynical and suspicious predisposition toward disbelief that assumes all new ideas and theories false until proven true by some insurmountable standard—a standard of proof that even most conventional medical beliefs would wither under. The well-known propensity for skeptics to repeatedly demand proof from those with new ideas not yet recognized by mainstream science is not, as one might assume, a sign of scientific rigor—it is a diversionary tactic used to put perceived opponents at a disadvantage. It is a sure sign of closed-minded disinterest. New ideas are perceived as a threat by skeptics because the implications of those ideas might require that they alter their basic assumptions.

The good news is that scientistic extremists stick out like sore thumbs and tend to discredit themselves by virtue of their outlandish claims and attitudes. The real danger lies with the impact of scientistic propaganda on the general population, which often assumes it to be true, and then parrots it back as gospel to others who, in turn, also believe it to be true. Skeptics understand this and make it their goal to influence society and public policy in order to advance their scientistic agenda. Even scientifically trained individuals can sometimes fall for the fantasy of scientific solutions to all human problems. Discerning people with and without scientific backgrounds recognize the limitations of science and are not fooled by skeptics' attempts to dominate the conversation. Mainstream thinking has thus become an indecipherably confused mix of sound ideas grounded in conventional science, overly optimistic projections of scientific success, and intellectual intolerance grounded in scientism. To the extent that unrealistic predictions and intolerance prevail, the greater the likelihood that undemocratic, autocratic, and even fascistic tendencies will be enabled to gain a foothold.

There is no better evidence of this muddled relationship with scientism than the current backlash against it coming from a truly odd mix of anti-government, anti-climate science, anti-evolution, anti-intellectual, and religious fundamentalist sentiment bursting forth in America today. The conservative right is justifiably concerned, but it is also quite confused, unable to discern the difference between conventional science and fundamentalist scientism. To conservatives it rightly feels like a threat to their values and religious autonomy. The conflict can be understood as an equal and opposite reaction from the right against the perceived encroachment of science and government on the left. It is first and foremost a fear-based reaction fueled by a lack of accurate information. What the right is actually reacting to is the dogmatic overreach of scientism and its growing influence in academia, medicine, industry, and government. Skepticism, therefore, simultaneously undermines the foundations of both science and religion. It presents itself as the epitome of fact-based reasoning while thinly veiling its faith-based scientistic agenda and monopolistic anti-religious designs on truth.

The influence of scientism causes many to defer to scientific authority on issues of public policy because it is assumed that scientists know best. The general public readily accepts favorable projections of scientific progress and tends to raise questions only when it is too late. Such is the case, for example, with rosy forecasts of the benefits that biotechnology will someday bring—prosperity, greater ease of living, and relief from human suffering. Many in the mainstream have faith in these yet-to-be-realized benefits of science and technology, but there are a growing number who do not share the same optimism. They view the world from a different frame of reference, from a broader, more holistic paradigm, and are deeply concerned with the unanticipated fallout that often comes with new technologies.

Make no mistake; the impact of scientism on medicine is growing. It is having a profound effect on patient care. Medicine has been stuck in a materialistic drugs-and-surgery mode for decades now. Medical academia is practically impervious to new ideas and this sad reality is reinforced by a slavish subservience to scientistic ideals of evidence and objectivity. No new-paradigm modality of treatment or theory of healing is capable of satisfying the narrow standards of old paradigm medical thought. Scientism is the new guardian of so-called evidence-based medicine. It ensures that potential paradigm-changing ideas will not be admitted into the medical mainstream. It virtually guarantees the continuation of the medical status quo.

The tendency towards orthodoxy is further reinforced by medicine's increasingly questionable relationships with corporate entities whose motives have little to do with health and healing and everything to do with profit margins. Although corporate influence is not a focal topic of this book, continuing down this corrupt road will ensure the eventual demise of medicine. However, that may not be a bad thing. Many are slowly coming to realize that corporate medicine does not have their best interests in mind. When critical mass is reached, like a school of fish that turns on a dime, the general public may suddenly decide to opt out, change course, and migrate to other forms of healing that better suit its needs.

When new medical modalities do make headway they often do so by compromising standards and acquiescing to mainstream medical control. Alternative medical organizations and modalities that choose to maintain their integrity often find themselves on the outside, isolated from mainstream opportunities and advantages. The history of the evolution of health care terminology helps to illustrate my point here.

The term *holistic medicine*, in my opinion, most accurately depicts the overall philosophy and intent of practitioners in the field. Over time, holistic medicine gained confidence and morphed into *alternative medicine*, perhaps to highlight the idea that it wanted to be recognized as a distinct alternative to conventional medicine. Some who resonated with principles of environmentalism preferred the term *natural medicine*. As Western culture enthusiastically embraced science as a whole, technological medicine rose to power and alternative or natural medicine found itself losing ground. It is no coincidence that the term currently preferred by mainstream medicine is *complementary medicine*, implying a subservient role that supplements conventional medicine. Feeling its oats again, the profession has recently tried to throw off its second-class citizenship by adopting the term *integrative medicine*, thereby indicating a cooperative relationship between conventional and holistic medicine as equal partners. My own moniker, *green medicine*, is intended as an all-inclusive version of integrative medicine that conveys an ecological awareness of the interconnectedness of body, heart, mind, and soul.

Interestingly, the historical name of a particular strain of American medicine from the late 1800's, *eclectic medicine*, reflected a spirit of inclusiveness and diversity that is sadly missing from the contemporary scene. Initiatives designed to establish peaceful relations with the mainstream usually fall on deaf ears. Mainstream medicine's scientistic refusal to acknowledge the undeniable value of alternative therapies has caused holistic medicine to rebrand itself under different names in the hope of achieving greater cultural acceptance. In my opinion, the profession should not compromise its principles in an attempt to appease conventional sensibilities. It should choose a name that is consistent with its own internal truth—and stick with it.

There is a big difference between those who pursue medical science with conscious awareness of the boundaries defined by its *a priori* assumptions, those who bring unconscious scientistic beliefs into the scientific arena, and those who imperialistically promote their scientistic agenda. The latter type intentionally resists medical innovation and seeks to define and control the body of medical knowledge and the practice of medicine. The inevitable outcome of scientistic control is medical authoritarianism, which, like all despotic impulses, finds its origin in fear of loss of control. Healthy skepticism coupled with open-minded curiosity is characteristic of a balanced attitude. Extreme skepticism is a pathological condition of mind that lends itself to scientific tyranny. The inescapable irony is that skeptical defenders of scientistic belief are the greatest promulgators of pseudoscience themselves.

Contrary to scientistic belief, medical science is not a fixed entity whose presuppositions, methods, and rules are written in stone. In fact, there is a good bit of evidence to support the notion that medical science evolves precisely because basic assumptions regarding the nature of life, health, illness, and disease also evolve. Our assumptions evolve by virtue of the lessons learned through practical experience. There is ample room for more than just one exclusive methodology to employ in the study of the magnificent, mysterious, and diverse phenomenon of human health.

Scientific knowledge is just one, albeit impressive, source of knowledge. But it is knowledge that is only skin deep. It is not capable of answering ethical or moral questions and it is mute regarding the metaphysical, spiritual, psychological, philosophical, and religious dimensions of human existence. Science has little of significance to say on matters of purpose and meaning, which, in my experience, tend to play the most important roles in the biggest health issues of all—the will to live, resistance to illness, and the capacity to heal.

Part III

Authentic Medical Science

CHAPTER 15

Toward a New Set of Assumptions

Wisdom begins with the recognition that our presuppositions are options that can be examined and replaced if found wanting.
  –Huston Smith, *Science and the Myth of Progress*

I regard consciousness as fundamental. I regard matter as derivative. We cannot get behind consciousness. Everything that we talk about, everything that we regard as existing, postulates consciousness.
  –Theoretical physicist, Max Plank

Assumptions made by a culture regarding the universe, humankind's place within it, and the nature of health and illness, will play a significant role in the way its medicine is practiced. Conventional medicine seems to think it is immune from philosophical prerequisites, choosing to ignore the topic completely, operating under the naïve assumption that metaphysics has nothing to do with science. Believing that it is possible to bypass the necessary preconditions, it instead focuses its energies on the so-called hard science of medicine.

Of course, this is precisely why medicine is so haphazard, prone to error, and known for its tendency to generate side effects and adverse events. One study contradicts another, which, in turn, contradicts another. We are led to believe that this is the price of progress—the trials and errors of an evolving science. I believe that the capricious changeability of medicine is a direct consequence of its lack of guiding principles. In the absence of those principles, it is left to tilt in the wind, changing course with each new study, unable to exercise the judgment and wisdom that comes from knowing oneself and one's core values.

Other than the largely unconscious –isms embodied by materialist science, medicine has very few if any philosophical principles that guide its activities. Although I am in disagreement with conventional medicine's basic presuppositions, it would serve the profession to become more acquainted with them. It would prevent a good number of mistakes, enable the profession to define its boundaries, and facilitate the creation of a much-needed line of demarcation between science and scientism.

The following is a list of axioms, the numbers of which correspond to the chapters in this book. This alternate set of assumptions represents both an answer to the –isms of conventional medicine and a preliminary formulation of some of the foundational principles of a new, more authentic approach to healing. These principles are not just logic-based, unfounded and ungrounded theoretical propositions. They derive from my own twenty-five years of holistic medical practice and the empirical experiences of thousands of practitioners of a variety of unconventional forms of medicine over the span of centuries.

These axioms are just a start; there is much more that can be said, but that is beyond the scope of this book. It should be remembered, too, that these differences in perspective are partially attributable to the natural differences between individuals due to their left, right, and mixed-brain modes of perceiving. It is not a question of right or wrong, or whose version of reality is more accurate. The final determinant of merit of any methodology must be the practical short and long-term outcomes that it is capable of producing.

1. Medical understanding changes with experience thereby giving birth to new presuppositions regarding the nature of health and healing.

Nowadays, the basic assumptions underpinning scientific method are assumed to be settled and true simply because they are associated with science. It only stands to reason that science must have carefully deliberated over its foundational principles at the outset, but this turns out not to be the case. Science, at its inception, was a response to authoritarian religious institutions and their cosmologies, thus imbuing science with an innate bias against the supernatural or any form of immaterial explanation for perceived phenomena. Over time the tables were turned; thousands of years of religious influence have been eclipsed by three centuries of highly successful scientific investigation of the material world. We in the West make the mistake, however, of assuming that the material success of science is evidence of the truth of its materialistic cosmological conception of the universe.

Medical science, in particular, is guilty of falling behind the scientific times by maintaining an essentially flat Earth perspective regarding human health. Although it has eagerly adopted diagnostic technologies courtesy of higher-level physics, it has failed to adopt the same principles to its understanding of illness and healing. Its mechanistic, materialistic, reductionist take on health is overly simplistic and surprisingly static. In spite of changes in scientific understanding, medical therapeutics has remained mired in a biochemical model, largely because it is so readily adaptable to an economy that thrives on the commodification of medical products.

Medical science should not be a fixed and rigid thing, and it certainly should never become a means by which to promote or defend a particular ideology or worldview. The presuppositions of science appear to have developed prematurely. By separating itself from the so-called superstitions of thousands of years of religious experience and spiritual wisdom it boxed itself into a three-dimensional prison comprised solely of matter and measurable energy. The principles of scientific method, therefore, were constructed in such a way as to reinforce a worldview that was already assumed to be the case before it undertook its investigations. As a consequence, science makes significant investments designed to maintain the status quo, which serve to protect the assumptions that define the parameters of the box that it has created for itself. At the same time it presumes to assert that all of existence fits into that same conceptually restricted box.

Medicine has neglected to update its metaphysical presuppositions in accordance with accumulating evidence. Instead, it denies the evidence, writing it off to anomaly or placebo affect or artifact, in the name of defending its unbending materialistic principles. While there is some adventurous research being conducted at the cutting edges of medical science, the practice of conventional medicine remains strictly old school—synthetic drugs, surgeries, and not much else.

History reveals that basic assumptions change as our understanding evolves. Holistic medicine is that faction of the medical profession that has refused to be kept in the box. It has broken free from dogma and lit out in search of less restrictive methods of investigation, more comprehensive philosophies of healing, and less toxic therapeutic interventions. The search has uncovered a number of remarkably viable alternatives, many of which have been around for hundreds or even thousands of years. Medicine not only lags behind advances represented by quantum theory, it also fails to incorporate new paradigm principles as elucidated by holistic systems of healing.

Holistic medicine is thought to be unscientific only when viewed from the limited perspective of conventional science. The truth is that holistic methodologies are no less scientific than those of mainstream medicine. They are simply different—and therein lies the rub—because, like religion, orthodox medicine contains a distinct streak of authoritarianism that has little tolerance for differences or dissent. Holistic principles are indisputably scientific when understood in their broader context defined by a revised set of basic assumptions. When the *–isms* of medicine are updated to reflect the wisdom inherent in alternative approaches to health and healing, patients become the therapeutic beneficiaries.

There is no doubt that science will continue to evolve. Its champions must possess the humility and savvy to incorporate new innovations into the day-to-day practices of the various scientific disciplines. Metaphysical presuppositions must not precede the evidence—it should be the other way around. The adoption of new assumptions should be based on the lessons learned through new observations and experiences. The best place to begin will be with a long overdue revision of the antiquated *–isms* of Western medicine so that they are made consistent with the fundamentals of holistic reality.

2. Non-physical phenomena are every bit as real as material ones, and both are crucial to the overall health equation.

The material nature of reality is not a factual certainty; it is a metaphysical assumption. Recall that even the concept of particulate matter, as we have known it, is under constant revision, thanks to the ever-changing face of contemporary physics. Evidence for the existence of non-physical reality, on the other hand, is overwhelming. That evidence encompasses all of human experience. The spectrum of consciousness is very broad indeed, broader than human experience is capable of documenting. Only in the counterintuitive minds of logicians could consciousness be called into question and made subservient to brain matter.

Dreams are nearly universal. Some are mundane and some are mysterious. Many dreams speak in archetypal language, with symbols so old and arcane that the dreamer in waking consciousness is incapable of comprehending them without assistance from scholarly texts and/or knowledge of mythology. Dreams can foreshadow the future and even warn of impending disease. Deceased loved ones often visit the living in dreams, and living loved ones often warn of their immanent departure. Some lucid dreamers are aware of themselves in their dreams and can influence the course of those dreams. There is nothing unusual about such events; they are ubiquitous and have been documented thousands of times over.

Further evidence of the broad spectrum of consciousness includes out-of-body and near-death experiences, widespread belief in reincarnation, vivid accounts of past lives, mental telepathy, precognition, clairvoyance, psychokinesis, and a wide range of spiritual experiences. Such phenomena are not, as scientists would like to believe, evidence of intellectual naiveté or mental instability. They are common events that cultural conditioning has convinced us to downplay, ignore, and even deny. The prevailing scientific mode of thinking makes us feel that there must be something wrong with us if we admit to similar experiences. We acknowledge them only to our closest friends and loved ones, leaving us to wonder alone if others are as crazy as we are.

This spectacular trove of non-ordinary consciousness could provide endless material for any truly curious scientist. But medicine chooses to look the other way, content with the way things are, and fearful of the implications for its carefully constructed fortress of rationality and objectivity. The paradox, here, is that medical science contributes to both mental and physical illness by refusing to acknowledge the personal truth of such diverse experiences. In doing so, it also misses out on endless opportunities to discern meaning and purpose in the lives of suffering individuals.

If all of the aforementioned is true, then why is it so hard to accept that mind can affect matter, that our thoughts and emotions can make us physically sick and make us well again? To take immaterial causation seriously would be to spark a revolution, altering everything that materialist mechanist medicine knows about mind-body relationships. Mere acknowledgement of this type of mind-body interaction, however, is not sufficient if it does not change the calculus of medicine, which will likely continue to seek material therapeutic solutions regardless of the actual etiology of illness. In this regard, there is a great deal to be learned from a variety of well-established holistic healing modalities.

Whether we can provide evidence as to the truth or accuracy of our subjective experiences should be neither here nor there. It is only relevant when we subscribe to the oddball parameters of materialist science, which demands physical proof to appease its pathological skepticism. Science distances itself from the mysteries of the mind in order to remain objective and in order to preserve the mechanics of its materialist philosophy. To do otherwise, it is believed, would be to betray science itself—such a strange position to take for a methodology that is touted as the most advanced means of acquiring knowledge that the world has ever known.

Consciousness, spiritual energy, life after death, and the preservation of the soul after the physical body has expired, are taken for granted by the world's religions. They cannot be discarded as mere superstitions. They are consensus impressions acquired from generations of human experience and accumulated spiritual wisdom. For medical science to arrive on the scene thousands of years later, expressing skepticism simply because such phenomena cannot be proven to the standards of randomized controlled trials is almost comical. It is not unlike a spoiled kid who bullies the other kids into playing by his rules, which are obviously rigged to turn the outcome in his favor. Consciousness is a self-evident truth that requires a great deal of twisted logic and rationalizing to refute.

Medicine's denial of the transcendent nature of consciousness is most evident in its frantic rush to forestall death at all costs. Even when all signs point to the completion of a person's time here on Earth, medicine, which has no understanding of and places no value on the concept of an afterlife, rarely allows a patient to leave the material plane with peace and dignity. Instead, it proposes numerous delay tactics in the form of heroic measures, all of which disrupt what should be a time of focused concentration on the higher purpose of the moment. Heroic measures are also a notorious drain on the emotional and financial resources of patients and their families. Western medicine values the preservation of physical life but is completely lost when it comes to otherworldly matters—spiritual matters towards which it displays a remarkable degree of insensitivity and skepticism.

The truth as told by holistic experience is that mind can affect body and body can affect mind. In fact, the mind-body-heart-soul is a seamless whole with no dividing lines. Explanatory gaps exist only in the minds of those who must have mechanistic reasons to justify that which is obvious to many. Some say matter and consciousness are on equal footing; others say consciousness is primary. There is no shortage of belief in the idea that all is mind, including the so-called physical universe. I say it does not matter for now, as long as we at least acknowledge the existence of the immaterial and the role that consciousness plays in health and illness. Once we agree to take that quantum leap forward, we can begin to debate the nuances of the issues at hand, and how science should or should not approach these issues. Until that day comes, conventional and holistic medicine will remain estranged cousins, unnecessarily suspicious of each other's motives.

3. While reductionist knowledge of local parts is indispensable, a comprehensive and complementary understanding of the whole provides a more reliable roadmap to overall patient health.

By definition, reductionism narrows one's perspective. It does so by focusing attention on one or a few parts to the exclusion of all others. Experience teaches us, however, that solutions based on just a few parameters often fall short of the mark. They either produce temporary results or generate unforeseen consequences that could not have been predicted without an understanding of the bigger picture. Due to the limited scope of its inquiry, medical reductionism often fails to perceive the impact of its actions on the greater whole.

This lack of concern for the larger picture is the most glaring weakness of conventional medicine. It tends to produce short-term results followed by relapses, side effects, complications, and longer-term chronic complaints. Such consequences are unavoidable when health is perceived only from medicine's fragmented, overly specialized, reductionist frame of reference. Conventional medical generalists are a disappearing breed, while specialists are a dime a dozen—and that is a recipe for serious medical dysfunction.

The larger problem, though, is that medicine's reductionist ways apply to far more than just the quarantine of body parts and organ systems from each other. Reductionism dissociates one part from another, the local manifestation of illness from the global health of the patient, and short from long-term consequences. On a more profound level, it provides justification for medicine to separate itself from human values. Once an ailing patient enters into the medical arena, his or her personal beliefs, individual sensitivities, moral concerns, and religious traditions tend to take a back seat to the clinical concerns of medical science. By the same token, clinical findings and stereotyped protocols are much more likely to drive decision making rather than the first-hand experiential feedback provided by patients themselves. In this sense, the needs of medical science often supersede the personal needs of patients.

Reductionist thinking allows an agency like the FDA to accept narrowly defined clinical trials as justification to approve drugs that later turn out to be harmful to patients. The parameters of randomized controlled trials are usually too limited to determine the real world impact of newly synthesized pharmaceuticals. Evidence of such harm ultimately takes the form of first-hand reports from patients and second-hand reports of physicians—only after flawed products have been authorized for widespread use.

The most problematic reductionist division of all is that of mind from body. Holistic experience repeatedly confirms the powerful impact that thought, belief, and emotion can have upon physical and mental illness. In my estimation, a significant majority of organic disease is either triggered by or maintained by psychological factors. What are we to make of a medical system that gives lip service, at best, to this reality?

Even when medicine reluctantly acknowledges the psychosomatic nature of a limited range of conditions it, nevertheless, continues to offer mostly biological solutions. The child struggling with a school phobia and headaches, therefore, receives the same old superficial painkiller approach, oftentimes leaving the deeper psychological source of distress unaddressed. Evidence of the profound nature of the mind-body connection and its role in health and illness is ubiquitous and, yet, medicine's reductionist orientation virtually prohibits it from detecting, let alone comprehending, that connection.

When properly understood, reductionism is both the polar opposite and complementary companion to holism. The remedy for medicine's lopsided left-brain reductionist emphasis, therefore, is not its replacement by holism but a synergistic alliance between the two. Most medical situations are best served when viewed from both perspectives. Well-established comprehensive holistic systems of healing like acupuncture, Ayurveda, Traditional Chinese Medicine, and homeopathy have a great deal to offer and, when combined with the reductionist understanding of Western medicine, can make an enormous difference in health care outcomes.

A balanced blend of reductionist knowledge and holistic wisdom is the ideal approach to health and healing. Figuratively speaking, after the parts are separated and studied in isolation they must be reconnected in order to understand their relationships. However, holism teaches us that the mere sum of the parts does not adequately represent the human organism as a whole. The mind-body-heart-soul in its entirety will always transcend all attempts to understand it in fragments. Reductionist solutions that are in sympathy with the principles of holistic healing are less likely to generate unanticipated consequences and more likely to promote long-term health and well-being.

4. There are a variety of relationships among medical events other than conventional mechanistic cause and effect.

Cause and effect is a type of relationship ascribed to so-called objective phenomena. Only concrete material forces, things, and events fulfill the arbitrary criteria of causation in science. Mechanism does not apply to mind, heart, emotion, spirit, or consciousness. I use the qualifier, "so-called," to emphasize the point that separation of subject and object is an artificial construct of left-brain thought not consistent with holistic reality. It is not possible, according to conventional science, for subjective events to have cause and effect relationships. This is mistakenly believed to be the case because the methods, tools, and presuppositions of science are not adequately equipped to detect or make sense of such relationships.

Since subjectivity finds no place in science, those unwilling to dismiss it have developed a variety of ways to understand it and its relationships to human events and physical phenomena. Perhaps the most famous is psychiatrist Carl Jung's notion of *synchronicity*. This phenomenon has also been called "meaningful coincidence," implying that there is more than just coincidence at work among phenomena that would otherwise have no discernable cause and effect relationship to connect them. Even when there is no objective connection, it is still possible to be connected in terms of meaning and significance.

When I dream of a turtle and then come upon one crossing the road a few days later, science casually dismisses this as coincidence. Now let's say that this occurs in the context of my studying American Indian creation myths, one of which involves the turtle's back as Mother Earth, the ground that supports all life forms. This confluence of events—my studies, dream, and encounter with turtle—are connected by a powerful sense of meaning that may have implications for my life, my future, and the choices I make.

Even though science fails to discern cause and effect among these types of events, this does not mean that it is not at play. It is just obscured by our limited capacity to perceive, our closed attitude toward the immaterial universe, that vast reality that lies outside the reach of the five senses and the linear, logical mind of science. When applied to medicine, this same limitation of perception can cause physicians to overlook meaningfully significant connections between medical events. This type of information can be just as useful as lab values and diagnostic imaging—if our minds are open to it.

I once saw a patient with Bell's Palsy—a form of facial paralysis—who reported a dream in which he tried but was unable to lift a child's swing set because it was made of lead. This led me to consider the possibility of prescribing *Plumbum metallicum*, a homeopathic medication prepared from lead. Although he had made very little prior progress for a number of months, a few doses of this medicine yielded notable improvement. Subjective information that would mean little from the perspective of conventional cause and effect provided the critical clue that led to the near complete resolution of this patient's palsy.

Because we cannot attribute cause and effect to this patient's dream and his recovery, we instead call it a synchronicity, a meaningful coincidence. We do so because we are incapable, with our limited scientific resources, of being able to perceive deeper forms of causation that escape the conventional laws of biology, chemistry, and physics. Those deeper forms of causation have been observed and studied by mystics, monks, poets, parapsychologists, and holistic healers throughout the ages. The topic is much too vast to cover here, but it is crucial for science to acknowledge that it does not have a corner on the market of understanding the relationships among phenomena, both objective and subjective.

Since medical science cannot explain relationships between mind and body in mechanistic terms, its solution is to essentially ignore the mind half of the problem. When other successful systems of healing propose a variety of intelligent theories and practical methods that incorporate both mind and body, medicine chooses to defend its *–isms* rather than open its mind to new possibilities. Two such systems are acupuncture and Ayurvedic medicine. Both are predicated on vitalist theories of health and illness. Mind and body are connected by a life force—chi in acupuncture, and prana in Ayurveda—a bioenergetic phenomenon that governs the organism as an organized intelligent whole.

Vitalism posits a life force distinct from conventionally defined biological, chemical, or physical forces. Since the life force cannot be directly observed or measured by the instruments of conventional science, it tends to be dismissed as nonsense. Nevertheless, there are those who claim to be able to sense the life force, some who can see it in the form of an aura, and systems of healing that employ practical methodologies that are capable of working in unison with the life force rather than against it. From the perspective of these modalities, cause and effect between mind and body, and vice versa, is an everyday reality. In fact, some, like myself, view mind and body as one and the same.

An additional type of relationship undervalued by medicine has to do with what science calls *association*. When a cause and effect relationship cannot be established between two events connected by time and/or space, they are said to be merely associated, as if to imply that association is meaningless. And, yet, pattern recognition is a powerful form of information gathering that can yield valuable healing knowledge. An assortment of phenomena unrelated by cause and effect may, nevertheless, occur together in time and/or space in a variety of patterns. Such patterns may contain important information that can be put to use in healing.

Homeopathic medicine is one such highly systematic methodology that makes use of pattern recognition in its therapeutic approach to illness. When a patient describes a constellation of objective and subjective complaints including, for example, left-sided migraines, dreams of snakes, feelings of jealousy and mistrust, and sleep apnea, the indicated treatment should be quite clear to any person trained in the rudiments of homeopathic practice. A different treatment would be indicated for a person who reports experiencing right-sided migraines, indigestion that acts up in the late afternoon, uncontrollable cravings for sweet foods, and a debilitating lack of self-confidence.

The first example is a recognized symptom pattern that corresponds to the homeopathic medicine, *Lachesis*, which is derived from a specific snake venom. The second example corresponds to *Lycopodium*, a medicine prepared from a plant commonly known as club moss. Each medicine, when administered appropriately, can result in relief from complaints associated with the unique symptom pattern that each is capable of causing, and therefore treating, in accordance with established homeopathic theory and clinical evidence.

That mind and body are connected in such a way as to indicate that many left sided migraines occur in individuals who are prone to jealousy and have dreamt of snakes is an indisputable phenomenon that any experienced homeopathic practitioner can confirm. Such clusters of symptoms, although not causally connected, are clearly connected from a radically empirical, phenomenological standpoint.

Perhaps the biggest failure of mechanistic thinking can be seen in the outcomes of therapeutic strategies that are based on proximate causation. Medicine's over-reliance upon proximate causation as a therapeutic indicator tends to result in short-term relief, unforeseen side effects and complications, and long-term suppression.

When a person receives a cortisone shot in an arthritic knee, it often brings remarkable relief from pain and inflammation. But when that same person returns weeks or months later to complain that the other knee is now flaring up, most physicians fail to connect the two events, writing it off to the expected progression of the arthritic condition. Most holistic practitioners consider the possibility that it represents a spread of the condition to a new location stemming from the suppressive influence of the original cortisone shot. It is a broader perspective that understands arthritis as a systemic disease. When the condition is approached in a local manner, taking into account only the painful knee, for which a cortisone shot is the standard treatment, the consequences for the whole person are oftentimes detrimental and unpredictable.

Local approaches to systemic conditions are a function of thinking in terms of proximate causation. Most of Western medicine operates on this principle. It is perhaps the main reason why modern pharmaceuticals are so dangerous and fraught with risk. They act locally in very specific ways while generating havoc throughout the human organism as a whole. While it is true that actual or ultimate causation is an ideal that is often not attainable, therapeutic interventions that take into account the broadest perspective possible usually yield the best results. As always, it is best to employ local treatment when a local solution is truly indicated, as in the surgical repair of a ruptured tendon, but fixing the part is not the same as healing the whole. A balanced perspective that considers both proximate and actual causation is the ideal approach to effective healing.

Another type of relationship among phenomena is that of *correspondence*. For example, a particular plant that has flowers that bloom only at night has been found in homeopathic practice to be a highly effective treatment for many children whose fears also bloom at night. I have personally used homeopathic *Datura stramonium* to successfully treat numerous children and adults who complain of nightmares, night terrors, fear of the dark, and a general fear of violence that intensifies during the night hours.

A similar correspondence exists between horse chestnut and the type of hemorrhoids that it can heal. The fruit of the chestnut is covered with sharp pointed spines and, in homeopathic practice, its therapeutic application is most effective in patients whose hemorrhoids are experienced as very sharp pains that feel like many little pointed sticks. These examples are reminiscent of the *doctrine of signatures*, a principle of correspondence long ago discarded by conventional medicine because it was judged to be superstition.

Even though such correspondences are very real, conventional science's need for uniformity, its need for principles that apply across the board, precludes it from recognizing phenomena of this nature. Conventional cause and effect is not the only game in town. There are a variety of relationships among phenomena that can be classified otherwise. They need to be understood for the important roles that they can serve in both diagnostics and therapeutics.

The powerful and ubiquitous cause and effect relationship between mind and body is woefully underestimated primarily because neither the human senses nor the instruments of science can detect that mysterious explanatory gap. Medicine will not acknowledge it without a concrete explanatory mechanism. It is a serious miscalculation to assume for this reason that the mind-body relationship is either non-existent or of minimal significance. Consciousness posits many more relations beyond the mere mechanics of matter. Thought, emotion, intuition, imagination, and will are a just few factors that can have a direct influence on the development of an illness and recovery from it. The future evolution of medical theory and practice will be based on our understanding of bioenergetics as demonstrated by mind-body relationships—and this understanding will emerge from the lessons taught by holistic methods of healing.

5. Rational conclusions are not made true by virtue of the soundness of their logic. They must be verified by personal experience and grounded in patient reality.

Much of the problem with contemporary science boils down to the fact that it has degraded the value of experience. An amazing testament to the influence of science is that it has so effectively eroded our belief in ourselves, our own judgments, and the lessons that come from our first-hand experiences. In medicine, it has taught doctors not to trust what their patients tell them. It encourages them to believe the medical literature more than their own observations in the clinic. This profound disconnect is the source of much sickness in the medical system, both figuratively and literally.

Recall how the Mayo Clinic concluded that, in their opinion, many current medical practices, assumed to be supported by evidence, turned out to be supported only by faith in the mechanisms of action of those practices. In other words, the rationales behind some practices were considered sufficient evidence to justify their use as treatments. Much of Western medicine is a product of this type of logic run amok. Uncorroborated theories hold sway by virtue of the materialistic, mechanistic, and reductionistic appeal of their assumptions and conclusions to left-brain thinkers.

The rational ruminations of the medical mind are pointless if they are not grounded in patient reality. It turns out that patient experience frequently contradicts that which makes logical medical sense. There is something wrong with a medical system that offers no recourse for a person who feels lousy and complains of a variety of symptoms even though he or she has been given a clean bill of health because the physical exam and lab results are within normal limits.

Feeling lousy is not sufficient evidence in the court of medical practice—period. Once all material markers of illness have been investigated, there is usually nothing more that medicine can do. It is simply not equipped to handle the subjective dimension of human illness.

In my experience, patients often report a variety of symptoms well before any scientifically defined evidence of disease becomes detectable. The strength of most holistic approaches is that they are designed to handle the subjective complaints of patients. They employ empirically based methodologies, not in the conventional sense, but in a way that makes meaningful sense and practical use of experiential phenomena.

A holistic perspective understands that subjective complaints can be the potential precursors to more concrete pathology.

Rational medicine's desire to confirm the rightness of its logic-based theories should never override the actual needs of patients. The logic that explains medical conditions is far less important than the experiential reality of patient outcomes. What makes logical sense is not often the same as what makes practical sense. When a patient complains of an unusual side-effect from a medication, the physician should not minimize it just because it is not documented in the medical literature. Patient reality should not be made to conform to medical theory; it should be the other way around.

Because medicine is unable to let go of its need to explain, it becomes distracted from the real goal, which is to truly heal by all means available while doing the least harm possible. Rational explanations of causation trap medicine into settling for short-term results based on faulty logic. Tortured rationales are used to justify the unfavorable risks versus benefits profiles of many interventions and treatments. Medicine gets so wrapped up in trying to defend its actions that it loses sight of what is important. The root of this dysfunction is medicine's excessively rational left-brain viewpoint and its attendant lack of common sense.

Just take one look at the stilted and lifeless nature of most research papers and you can see that medicine is out of touch with its humanity. The technical jargon, objectified acronyms, preoccupation with method, statistical technicalities, and tentative, non-committal conclusions of most research papers render them so dense as to be impenetrable and impractical. Medicine's rejection of personal experience and the subjective nature of health and illness come at a high price. In trying so hard to be rational, it loses its relevance to real people with real problems.

It is time for medicine to reevaluate its priorities and return personal experience to its rightful place in the hierarchy of knowledge. Without experience there can be no science—or anything else, for that matter. To arbitrarily dismiss experience as unreliably subjective is a serious misjudgment grounded in left-brain bias. It would be just as easy, even easier, to point out the pitfalls and falsehoods inherent in abstract thinking. Experience is primary; it precedes the filter of mental deliberation. As such, it represents a more direct apprehension of things as they are, before they have been twisted and reformulated into the theories of science and practices of learned professionals.

The need to explain is, at bottom, a function of pride and ego, which is rarely shy about asserting the rightness of its rationales. All other forms of knowledge are thought to be flawed because they do not provide the certainty that logical proofs supposedly do. It is one thing to possess skills of argument and debate; it is another to possess the knowledge and wisdom that can facilitate the healing of humankind.

Patients should not be fooled by empty displays of scientific knowledge, especially when their needs are not being met. It is time to reject the contention of twenty-first century rational man that our personal experiences are meaningless anecdotes. Our experiences often hold the keys to our salvation, healing, and return to health. Because science makes no accommodation for personal experience does not mean that it is not of great value. *Anecdote* is a term co-opted by science for the purpose of keeping laypersons submissive and in line.

The same applies to physicians. They must learn once again to embrace the knowledge that comes from clinical experience. Rational knowledge is not the only form of knowledge. First-hand reports of patients and case histories provided by physicians are more directly relevant to patient care than many of the findings gleaned from abstract research trials. In truth, our patients have more to teach us than all other sources of knowledge combined.

Science is based on observation and experimentation, which is fancy way of saying trial and error. Rational explanation comes later, by virtue of the lessons learned from experience. Medicine must loosen its grip on its ideological need to provide rational explanations and begin to focus instead on real solutions. Many health-related problems are just not explainable in rational terms. Rational conclusions must be verified by experience and grounded in patient reality in order to pass the tests of practical usefulness, reliability, and holistic integrity. The art and science of medicine requires a fine tuned balance of reasoned analysis guided by the wisdom that comes from practical experience.

6. It is an illusion to believe that subjective phenomena can be separated from so-called objective reality and excluded from scientific and medical deliberation.

The entire thrust of medical objectivism is to force reality to conform to its science-based worldview. It does so by cleaving the world in half, leaving some of the most crucial elements of healing behind on the chopping block. But objectivity alone is an illusion. No matter how hard science tries, subjectivity cannot be ignored.

Without mind, spirit, soul, will, intention, intuition, imagination, myth making, art, magic, symbolism, meaning, and personal experience there is no human life. Any medical intervention that fails to account for these factors is doomed to mediocrity from the start. To reduce issues of human health to objectively measurable terms is to treat people as if they are mechanical automatons.

Reality, it turns out, is holistic. How could it be otherwise? Anything short of the whole gives us a fragmented, condensed, or truncated view of holistic reality. Objectivism cannot help but distort that reality. Objective reality is not a natural reality. It is an artificial construct of left-brain analytical thought, lacking in the qualities that make human life human. Objective medical science does not and never will be able to describe reality as a whole. It deals only with very small and limited parts of it.

The subjective is not rendered inferior simply because it cannot be confirmed by outside observers. It is only inferior in the estimation of men of science. It is counterintuitive and patently absurd to think that human consciousness is an illusion. Shall we continue to ignore consciousness because objectivist science has defined it out of existence? What could be more relevant to human health than the study of consciousness? Or shall we look the other way, pretending that human events, emotional circumstances, temperaments, beliefs, and states of mind have little to do with illness and healing?

It is admirable that physicians, clinicians, and healers strive to be as objective as possible—but only up to a point, and under the right circumstances. Taken to an extreme, objectivity is nothing more than a misguided value imposed upon our medical system by the supposedly value-free mind of science. Objectivity is a value of science; subjectivity is an experiential reality shared by all of humanity. Excessive detachment from the subjective realities and actual problems of patients is counterproductive to genuine healing.

The great paradox is that exclusion of the subjective makes it impossible to be objective. How can one be unbiased when one's worldview prohibits the fair consideration of human experience and consciousness? There is much room for improvement and a great deal can be learned from those who have made a career of studying consciousness. Alan Wallace offers just one among many such possibilities:

> *In their explorations of the mind, contemplative traditions have used a different definition of objectivity. Since their "experiments" use a first-person, subjective approach, the notion of distance or isolation proposed in the scientific view of objectivity is irrelevant. Rather, many contemplatives believe that they can get a clear picture of mental phenomena by the skillful use of meditation techniques. The basis for that skill—using another, somewhat different, connotation of the word objective—is to be unbiased. One can be subjectively involved with phenomena—close, rather than isolated—without introducing distortions. ...One then uses one's enhanced powers of mental perception to learn to distinguish between the phenomena that are presented to the senses...and the conceptual superimpositions that one under normal circumstances compulsively projects upon those phenomena.*[51]

Not surprisingly, it is the scientific mind that compulsively projects its conceptual limitations upon the phenomena that it observes. In medicine, those limitations are defined by the *–isms* that have been outlined in this book.

The objectification of disease into clean-cut diagnostic categories is another way in which medicine does a serious disservice to the realities of suffering individuals—individuals who frequently fall through the cracks because their ailments do not meet the cookie-cutter standards of disease classification. Most holistic practitioners can testify to this because they are the ones who those individuals turn to when the medical system fails them. Treatment of an illness as an objective stereotype will always have a low probability of success because it is nothing more than an abstraction from holistic reality, an abstraction that may or may not correspond to the actual illness of the patient.

Treating disease as an object also creates the illusion that it is permanently eliminated once it has been subdued. Oftentimes, medicine equates suppression with cure—out of sight, out of mind. By this standard, the elimination of eczema with topical cortisone and the eradication of an ear infection with an antibiotic represent successful cures, regardless of whether the eczema returns a year later or the ear infection metastasizes into pneumonia a month later. But all holistic healers know that health and illness are dynamic states, always in flux and subject to change. Healing is an ongoing process, not a static state represented by a moment in time, not a commodity to be purchased from the medical establishment, and not a product to be applied to the sick.

Subjectivity goes hand-in-hand with personal, first-hand experience. It, too, must be elevated to its rightful place alongside objectivity. It is the one missing element capable of restoring purpose and meaning to the medical experience. While the purpose of objectivity should be to ensure evenhanded open-mindedness, careful consideration of the subjective dimension of illness guarantees that the real and diverse health problems of actual persons will be addressed in an effective manner consistent with holistic principles.

7. Medical empiricism must be revised to include consciousness and all types of subjective experience.

The empiricism of Western science allows only for the objective correlates of sensory experience. It excludes the sensations themselves, and the entire realm of extra-sensory experience. That realm, the realm of consciousness, is not directly accessible to objective science. Subjective experiential phenomena such as thought, emotion, intuition, intention, insight, epiphany, imagination, symbolism, love, and meaning are best judged from a first-person perspective.

Experience is primary. Rational analysis comes later. Experience is pre-cognitive; we think about our experiences only after they have been experienced. Experience is the raw data that constitutes the subject of science. Science subjects the world of experienced phenomena to its objective methods of rational analysis. Scientists construe experience; they interpret experience through that unique filter called science. Even from a scientific perspective, it is possible to arrive at multiple interpretations of experience.

There are many who have no problem with subjecting the human mind and consciousness to scientific scrutiny. In fact, it has been done for hundreds if not thousands of years. Buddhist monks, Hindu swamis, Christian mystics, and contemplatives of all stripes have made rigorous disciplines out of studying the inner workings of the mind, consciousness, and their relationships to creation. Sadly, contemporary science does not acknowledge this vast tradition of experiential empiricism.

Alan Wallace has spent a great deal of time exploring the implications of the study of consciousness through introspective disciplines like yoga, meditation, chant, and prayer. The physicist-monk laments science's prohibition on the lessons learned through introspective practices:

> ... *many Hindu, Sufi, and Buddhist traditions have made logical and systematic explorations of the mind resulting in experimental tools and techniques (of a first- and second-person nature), possess a critical literature, and sometimes incorporate academic debate. These are all hallmarks of an empirical tradition. ... As long as science accepts its own brand of empiricism, its own definition of objectivity, and its own tools of measurement as the only valid ones,*

it will miss a golden opportunity to impartially evaluate what may
be a treasury of knowledge of the mind.[52]

It was William James who proposed a radical empiricism that did not
shy away from subjective experience. He believed experience to be the
most elementary feature of reality. After all, what is science but a particular method of interpreting experienced phenomena? James made it
clear that nothing was to be excluded.

To be radical, an empiricism must neither admit into its construc-
tions any element that is not directly experienced, nor exclude from
them any element that is directly experienced.[53]

Inclusion of directly experienced phenomena in the definition of empiricism would change the way that science and medicine are practiced.
It would mark an important step in the evolution of medicine toward
holism. Here, Alan Wallace makes the case for the magnitude of James'
contribution to genuine holistic empiricism:

Like the Copernican shift from a geocentric to a heliocentric view
of the solar system, the shift from scientific materialism to radical
empiricism entails a shift from a matter-centered concept of reality
to a holistic view of mental and physical phenomena as dependently
related events.[54]

The first step in healing is to acknowledge the fundamental truth of
a suffering individual's experiential reality. When medicine invalidates
experience it undermines the healing process. By remaining open to all
possibilities, we serve the highest good of our patients. New experiences
can lead to new assumptions regarding the nature of health, illness, and
the healing process. This is how science, medicine, and healing evolve.

By contrast, conventional physicians are supposed to rely on the "literature" to determine proper protocol. If they were to truly pay closer
attention to their own clinical experiences, they would be compelled to
reevaluate what some of those studies are telling them. Medical empiricism is an elaborate rationalization that objectifies experience, ensures

methodological adherence to the *–isms* of medicine, and prevents physicians from thinking freely.

Conventional medical science wants to have its cake and eat it too. It claims to be empirical while simultaneously taking alternative therapies to task for being empirically based, which really turns out to be code for not conforming to conventional medical theory. If it is to withstand critical examination, medicine must articulate a more consistent definition of empiricism. In its revised form, it would be a crucial component of a renewed philosophy of medicine and healing, one that honors holistic empiricism but also accommodates the best of conventional medical theory. Alan Wallace sums it up nicely:

> *The time has come to join hands, uniting the knowledge and power of science and technology with the wisdom and compassion that characterize genuine spiritual realization. With the abandonment of all dogmas, let us return to a true spirit of empiricism, embracing the whole of nature, one that includes the objective world of science, the mind, and spirituality, and in so doing bring about a new renaissance that will heal our troubled world.*[55]

Empiricism and theory are not mutually exclusive. It should not boil down to a choice between observation and abstraction. It is not possible to have adequate knowledge of the world unless we account for both personal experience and the ability to reason. The Cartesian notion that rational analysis will always eclipse the lessons learned from direct clinical experience must be rejected. Reestablishing the legitimacy of experiential authority will go a long way toward restoring equilibrium to Western culture's lopsidedly rationalist outlook.

Honoring and respecting the first-hand experiences of patients is a critical prerequisite to the healing process. Physicians, too, must rediscover the powerful lessons that can be learned from their clinical experiences. Experience and theory are complementary aspects of all truly effective systems of healing. The reputation of personal experience must be rehabilitated. Subjectivity must be returned to its rightful place in medical philosophy if the art and science of medicine is to continue to evolve for the benefit of humankind.

8. Diversity, plurality, and individual choice should supersede the medical desire for conformity and its quest for the one correct approach to health and healing.

It is one thing for science to seek commonalities among the phenomena that it studies. It is another to elevate uniformity to a principle that defines science itself. Conformity and universality are not prerequisite goals for conducting medical science and are definitely not intrinsic features of scientific method. They are subjective values imposed upon science by those who teach, promote, and practice it.

Medicine is a particularly conformist milieu. All new ideas must run the gauntlet of *–isms* in order to ensure compliance with established belief. Physicians are discouraged from and even ridiculed for expressing dissenting opinions. They risk losing their medical licenses when they dare to practice outside the bounds of accepted care, regardless of whether their therapeutic methods are effective or not.

Holistic practitioners commonly endure the social and psychological stress that comes from practicing a standard of care that does not meet with approval from the medical mainstream. My experience has been no different, regardless of the fact that the homeopathic medicines that I prescribe have a two hundred year history and are FDA approved and regulated. Mainstream medical groupthink is so prevalent that few are even aware of the legitimate legal standing of homeopathy.

It is a truly sad commentary on the state of Western science when we realize that new ideas that do not satisfy the presuppositions of conventional medical theory are not welcomed. They are simply turned aside without a second thought, rejected by the guardians of conformity with barely an inkling of curiosity about what those new ideas may signify or whether they can be of service to suffering humanity. Unless a serious revolution in thinking occurs, we will be locked in to a corporate standard of care for a long time to come. If the guardians of purity continue to prevail, with few exceptions, pharmaceuticals, surgeries, and diagnostic technologies are all that will be permitted.

It is an error of judgment to think that diagnostic labels and their corresponding therapies can be generalized and applied across the board. The concept of individualized treatment and the reality that all patients with asthma, for example, are unique, has no place in this kind of medical environment. Likewise, it is a mistake to believe that research results that apply only to a few should be rejected as statistically insignificant while research that shows applicability to many should be highly valued. This notion is incompatible with the empirical evidence represented by holistic reality.

Although it would make life a whole lot easier, unfortunately, one size does not fit all. There is rarely ever one single correct way. The complexity, diversity, and unique circumstances of human illness preclude any straightforward formulaic approach to healing. When medicine imposes its will upon this diversity in an attempt to create order, the result is a multitude of errors—including side effects, complications, discredited research studies, recalled medical products, and dissatisfied patients. It is not a matter of improving existing practices until they yield better results. More successful outcomes will become possible only when basic assumptions regarding health, illness, and healing are changed.

The enforcement of a code of –isms on the theory and practice of medicine stems from a deep need for medical uniformity. The same need may be the biggest factor standing in the way of medical freedom for patients and practitioners alike. The medical system preaches compliance, fosters dependency, discourages innovation, and shuns freethinking. There is a reason that philosophy of medicine is a dead discipline—no one within the system is truly free to question the fundamental assumptions that inform Western medical practices. To do so is to invite accusations of unscientific heresy.

Nevertheless, there is a tremendous need for freedom of medical choice. It would enable physicians and patients to explore their medical options without judgment and without the threat of legal retribution. To persecute a person for his or her medical choices not only runs counter to democratic principles, it also implies that there is one correct way that will virtually guarantee the best medical outcome. While medicine clings to the illusion of its own certainty, there is never any guarantee of success. There is always plenty of gray area.

The way forward points to patient autonomy, freedom to choose, a diversity of viable methods, and a humble acknowledgment on medicine's part that it does not always have the answers. Medical science must begin to examine its misguided need for conformity. It is time to relinquish the authoritarian tendencies and dogmatic views that have squelched medical creativity and innovation for the past hundred or more years.

Of course, the flip side of the issue involves patient responsibility. The freedom to choose entails an acknowledgment of the risks involved. Human health is not a hard science; it is fraught with uncertainty and this must be understood by all involved. When something goes awry, it should not be reason to find fault as much as it should be an opportunity to learn from mistakes. There is plenty of blame to go around. Let those who would accuse unconventional therapies of being dangerous first shine a light on the tremendous risks involved in many conventional medical practices.

Plurality, diversity, and individual choice combined with humility, discernment, and personal responsibility is the only viable answer to our current closed-minded medical state of affairs. We may have a long way to go, but it is the only way to go.

9. Beneath the surface impression of dualistic separation lies the reality of an all-encompassing unified whole.

The concept of dualism and all of its implications become all the more convincing the closer we come to a purely materialistic conception of existence—hence, conventional science's love affair with dualism. The more room we make for consciousness, spirit, and the ineffable, the less sense dualism makes. Of course, dualism would have us split physical and spiritual existence into two distinct, unrelated categories. The naiveté of materialism permits this split because it does not believe in the reality of spirit and wants to distance itself as far as possible from any association with the immaterial.

A more mature perspective reflects a deeper and more inclusive holistic understanding of existence. From this perspective, it is impossible to ignore the influences that psyche, spirit, and energy can have on matter. From this perspective, the practice of physical medicine in isolation from the whole becomes an exercise in futility. The more fully we embrace the actuality of a holistic universe and all that this implies, the more likely we will be able to create a truly effective green system of healing that addresses body, heart, mind, spirit, community, and environment.

The polarities that we perceive are but the surface appearances of an underlying unified whole that has been recognized for millennia by the world's great spiritual traditions and that is now in the process of being reaffirmed by modern physics. This by no means makes the principle of polarity irrelevant. Polarities are very real and must be accounted for.

A suppressive drug that eliminates a set of symptoms is likely, sooner or later, to generate a physiological backlash, usually in the form of what tends to be perceived as a new and unrelated manifestation of illness. An overly lenient childrearing attitude neglects to take into account a young person's need to develop a sense of self-discipline, which can lead to a lifetime of psychological impulse control issues. A warlike approach to killing microbes fails to take into account our mutually beneficial symbiotic relationships to those same microbial neighbors—while at the same time ignoring the principles of natural selection that yield more virulent bugs as a consequence. In each particular case, an action that takes into account only one polar aspect of the problem at hand, not unexpectedly generates an overall imbalance, which manifests as a seemingly new but related health issue.

We live in a holistic universe where our shortsighted actions oftentimes have unforeseen consequences. Modern medicine has developed a powerful and efficient capacity for scientific tunnel vision, which enables it to ruthlessly accomplish its goals with little regard for the bigger picture. It defines those goals in strict material terms, failing to account for the impact that its actions can have upon the hidden totality. Although a particular action may be justifiably aimed at a local manifestation of an illness, we must always remain cognizant of the effects that it can have on the greater whole.

Dualism is an illusion in the sense that attention to a part can give a false sense of confidence that one is addressing the true problem. In truth, most health problems are not local. Most health issues are complex, multifactorial, energetic disturbances that are not visible to the eye except in terms of their local physical manifestations within the human body. It should come as no surprise, therefore, when a cortisone injection for an arthritic joint leads to the subsequent inflammation of a new joint previously untouched by arthritis. Arthritis is a systemic illness. Local approaches can only be palliative and oftentimes lead to an overall worsening of health.

There are many great thinkers, spiritual figures, and religious cosmologies that take consciousness to be the foundational principle of the manifest universe. Only modern science conceptualizes existence in the reverse order, whereby the physical comes first with all else emerging from the gradual evolution of matter. In typically linear fashion, the cosmos is perceived as a purely physical phenomenon that began with a bang and has been expanding ever since. Western religions tend to follow suit in that they are similarly linear and dualistic; we are born, we die, our bodies return to dust, our souls return to God, the end.

More sophisticated cosmologies incorporate cyclical principles of life, death, and rebirth. According to the *as above, so below* principle, therefore, the universe expands over eons of time, contracts back into the void, and then expands again, each time bringing with it evolutionary implications not just for humankind, but for the universe as a whole. In similar manner, souls are incarnated and reincarnated repeatedly until certain lessons are learned, after which the soul is ultimately released back into spirit.

Some interpret this as an equally dualistic representation of spirit and matter. Others counter that this is a mistaken notion because, as is

the case with all polarities, there is always a continuous spectrum between the two. In this instance, we are talking about a spectrum that
moves from concentrated spirit on the one end to concentrated matter on
the other. According to this perspective, rocks are not completely inert
because they represent a rudimentary form of energy or consciousness
made manifest on the dense material plane. Humans, by contrast, are
significantly more imbued with consciousness. A disembodied spirit vi-
brates on a higher plane and represents a higher form of consciousness.

When we place human health, illness, and its cure in this metaphysical context it highlights the conceptual and practical limitations of conventional material medicine. Dualism and reductionism can be misleading when understood in isolation without the glue that holds everything
together, the unified background of holistic interconnectedness. The implications are not just high-minded concepts for philosophical minds;
they have important real world consequences. It cuts to the heart of the
differences between orthodox medicine and holistic healing.

The world is full of polarities, but polarities always constitute two
aspects of the same whole. I am not endorsing a dualistic competition
between orthodoxy and holism, because holism implies the one whole
that embraces all. There are times for conventional approaches, times for
holistic measures, and times for both. Each has its strengths and weaknesses. It is possible to begin with the assumption that there are no actual divisions and still talk about local phenomena and isolated parts for
the sake of practical convenience. In the same vein, right and left brain
are complementary hemispheres that always work together synergistically. In combination, reductionism and holism are significantly more
powerful than when pitted against each other in a dualistic struggle of
polar opposites.

The mind-body problem is only a problem when one subscribes to
the many *-isms* of Western medicine. Dualism is an inferior position
that assumes a false dichotomy; it sees separation and division everywhere it turns. There is much to be learned from Eastern perspectives,
which tend to be sympathetic to and more compatible with holistic reality. Medicine can no longer ignore consciousness while tending only

to the physical body. Everything organic and inorganic is imbued with varying degrees of consciousness. All of existence is sacred and must be treated as such. If not, our capacity for healing will suffer and fall short of the mark.

10-11. There are forms of knowledge beyond rational knowledge that can contribute to greater understanding of the whole patient.

When Western philosophy and conventional science split the world in half they declared, in essence, that there is only one legitimate type of knowledge. The only way to validate their dualistic take on existence was to demote the subjective to an inferior, if not non-existent, status. They tried to solve the dualistic dilemma by pretending that one half of the problem was of no relevance to scientific or philosophic concerns. By default, objective knowledge became the only real knowledge. The entire edifice of Western science has been built upon this faulty premise, and it is still believed by many to be an unassailable truth.

Since this schematic of the natural world appears from a left-brain mode of perception to be virtually self-evident, it is rarely ever challenged. When the scientific worldview is questioned by the likes of artists, musicians, mystics, and metaphysicians, their reservations are easily deflected by the self-serving claims of scientists who argue that they could not possibly understand because they are not trained scientists. Most fall for this circular argument, regardless of its baseless unscientific nature, and are cowed into compliance. They fail to comprehend that one need not be a scientist in order to understand the scientific perspective as a whole.

Thus, the information provided by left-brain dominant thinking is all that really matters in science and in medicine. This is the same information that has been quantified, organized, sanitized, and homogenized for consumption by the abstractions of rational analytic thought. The same information represents second, third, fourth, and fifth-hand knowledge, multiple steps removed from direct empirical and experiential knowledge. In summation, it is the knowledge birthed from Western dualism.

It has already been a full century since quantum theory determined that object and subject are inseparable. The so-called objective observer cannot separate him or herself from the subject of investigation. Holistic reality ensures that scientists cannot conduct research without influencing their own experiments. Author and philosopher, Ken Wilber, in his classic book, *The Spectrum of Consciousness*, explains:

> In other words, when the universe is severed into a subject vs. an object, into one state which sees vs. one state which is seen, something always gets left out. In this condition, the universe "will always partially elude itself." No observing system can observe itself observing. The seer cannot see itself seeing. Every eye has a blind spot. And it is for precisely this reason that at the basis of all such dualistic attempts we find only: Uncertainty, Incompleteness![56]

Dualism is just a way of looking at things. It is a Western, left-brain, scientific mode of perceiving. This mode of perception, when taken to be the only valid avenue to knowledge, has enormous philosophical and practical consequences:

> So just as front and back are simply two different ways of viewing one body, so subject and object, psyche and soma, energy and matter are but two ways of approaching one reality. Not to realize this, and to set the "opposites" against one another while trying to figure out which is "really" real—this is to condemn oneself to the perpetual and chronic frustration of trying to solve a nonsensical problem.[57]

Wilber notes that all attempts to capture reality in words are doomed to failure. All words are second-hand symbolic placeholders for something else. The ideas produced by rational analysis are likewise abstract expressions that distance one from the phenomena that they are attempting to describe. Medical constructs, too, tend to be quite abstract and elaborate, and are a far cry from the first-hand experiences of the illnesses that they refer to. Rational knowledge, by its very nature, is a dualistic abstraction. However, as Wilber explains, there is another mode of perception that can provide a different but very valuable form of theoretical and practical knowledge:

> If reality is inexpressible, it is nevertheless experienceable. But since this experience of the real world is obscured by our concepts about it, and since these concepts rest on the split between the subject that knows vs. the concepts that are known, all of these traditions emphatically announce that Reality can only be experienced non-dually, without the gap between the knower and the known, for in this manner alone is the universe not delivered up to illusion.[58]

When the forbidden half of dualistic prohibition is examined in the light of day, we are forced to reconsider a number of assumptions that, heretofore, have been considered sacred and off limits to the queries of philosophers of science, philosophers of medicine, and scientists themselves. That other half, of course, is the world of immediate experience. Experience is the lens through which all human beings perceive all of existence, including the world of science. It is the lens that yields direct, first-hand, subjective knowledge. Experiential knowledge turns out to be no less a form of knowledge than is rational knowledge. It is possible to experience holistic reality non-dually—before the mind steps in to divide it into the object of experience and the subject who experienced it. Experiential knowledge entails more than just the materially dumbed-down realm of scientific empiricism. It is radically empirical knowledge that acknowledges all phenomena, both objective (as defined by science) and subjective.

When the blinders fall away, it becomes clear to many that dualism is an illusion perpetuated by the way we have been conditioned to think about life, existence, and the world we live in. The appropriate response to this revelation is not, as fundamentalist science has done, to attempt to invalidate and overthrow the opposing camp. That would be to fall back into the same trap of either-or, black and white, right or wrong dualism. The correct response is to see dualism as it truly is, as a way of looking at things, as a means of achieving certain practical aims. Dualism is a worldview that engenders a scientific method with limitations that must be consciously assimilated in order to avoid the pitfalls that come with those scientific blinders. It is when medical science acts without awareness of those limitations that it does the most harm.

Contained within the vast spectrum of consciousness are myriad phenomena all of which may have relationships with and an appreciable impact upon health, illness, and healing. Given our current understanding of the human psyche, it would be indefensible not to factor experience, consciousness, and the subjective into our medical calculations. Direct, non-dual knowledge takes many forms that cannot be duplicated in abstract rational or verbal terms. Imagination, emotion, intuition, epiphany, revelation, dreams, psychic phenomena, meaning, will, and intent, are just a few of the experiential contents of consciousness that, when rendered into words, lose much of their potency and original purpose.

Science may not be able to measure or quantify non-dual experiential knowledge, but it can acknowledge the role that it plays and the difference that it makes. This is not a simple call to respect patient experience for the betterment of the doctor-patient relationship. Experiential knowledge can help guide physicians in decision-making, diagnostically, therapeutically, and prognostically. It can give both doctor and patient the personal confidence and confirmatory evidence that can help validate certain medical strategies and lead them to question other strategies.

The real experts who have some understanding of the structure and dynamics of consciousness are not to be found in the medical field. They are the psychics, parapsychologists, and students of consciousness. They are the mystics, monks, swamis, and spiritual devotees. They are the alchemists, astrologers, and intuitives. And they include holistic practitioners, therapists, and healers of many persuasions. They are the ones who have devoted their lives to exploring the blind spot that rational medicine is incapable of perceiving without outside help.

The hypertrophy of the rational function parallels the historical rise of science in Western societies. Both are a consequence of overestimating the value of a left-brain mode of thinking. By limiting the knowledge base to one particular way of knowing, we are undermining the very essence of science, which is intended to be an open-ended tool for exploration free from exclusionary bias. A synergistic meeting of dualistic rational and non-dualist holistic modes would go a long way toward serving those who are in need of genuine healing.

12. The hierarchy of medical evidence is badly skewed and needs to be revised to account for personal experience and the subjective dimension of human health.

Some who subscribe to conventional scientific method as the only reliable source of knowledge also believe that whatever cannot be confirmed by this method should be assumed to be false until proven otherwise. In other words, if science has not yet proven it, then it is not true. This may seem like an overstatement but it is the primary rationale for rejecting outright most forms of alternative medicine and holistic healing. Until they can be proven according to the standards of EBM they are assumed to be untrue.

It is akin to being found guilty until proven innocent, but the rules of the court that could exonerate you are fixed in such a way as to make the finding of your innocence highly unlikely. For this reason, many alternative medical modalities are derogatorily labeled "junk science," or "pseudoscience." Such labels themselves are indicative of a closed-minded, unscientific attitude. When belief in science as the only dependable source of knowledge reaches this level of intolerance, it becomes *scientism*. Those who pass unfounded and unscientific judgment on alternative therapies fail to understand that most holistic modalities are grounded in different assumptions regarding the nature of health and illness. As such, they often employ different approaches to scientific method.

Mainstream medicine remains largely uneducated regarding a multitude of additional forms of evidence, each of which may be of value in the overall holistic assessment of a given situation. First-hand patient reports and physicians' impressions of those patients are of most relevance. Individual case studies and case studies involving small groups of patients are also quite valuable. Amazingly, these patient and doctor reports are what many medical academics disparagingly call *anecdotal* evidence. It is time to recognize the insidious nature of this term. It is nothing more than an institutional form of bigotry and its use in scientific settings should be rejected.

Outcomes-based research is concerned more with actual benefits as they apply to patients rather than the efficacy of a particular drug in the sterile setting of an RCT. Surveys can supply information that is directly relevant to patients' levels of satisfaction and sense of well-being. Comparative effectiveness studies compare multiple interventions, and in this sense are somewhat more holistic than RCTs. They also gauge effectiveness rather than just efficacy. Even epidemiologic observational studies can provide real world information about real patients in ways that RCTs cannot.

I firmly believe that we must turn the pyramid of evidence on its head. Truly evidence-based medical practice should rely on a much broader foundation of knowledge and information. First-hand reports from patients and second-hand observations by doctors of their patients should be given the predominance of weight when judging the value and effectiveness of any therapeutic intervention. Randomized controlled trials belong near the bottom of the hierarchy of evidence.

An RCT is a highly abstract, quantitative, excessively rational, mechanistic, and reductionist construct of left-brain thinking that, in addition to being highly biased in favor of a conventional medical perspective, has little holistic bearing on outcomes involving real patients and their actual health problems. Statistical outcomes are abstractions from holistic reality that can be used to justify any number of deceptive or erroneous conclusions. RCTs, rather than eliminating bias, practically guarantee that bias is structurally encoded into scientific methodology.

It should be stated that I am not claiming that patients should be the judge, for example, of whether their thyroid lab values are normal or not—that requires the assistance of a physician. But patient feedback can be crucial in determining if there is a thyroid problem to begin with, and whether treatment is providing effective relief. Even after lab values are restored to normal, patients can continue to feel miserable and unwell, and this should be of concern to any dedicated healer.

The peer review process should not be overlooked since it, too, is susceptible to abuse on multiple levels. Reviewers should not be chosen on the basis of their allegiance to the presuppositional foundations of medical theory or conventional standards of medical practice. There should also be a mechanism by which reviewers themselves are evaluated. Pre-publication peer review is particularly biased, especially when the identities of authors and their affiliations are made known. One solution is to make all authors and affiliations anonymous during the review process. Another is to wait until after publication to subject authors' papers to peer review. The purpose of peer review is to provide constructive feedback to authors, not to censor their work in the name of scientific purity.

It is time for medical practitioners to reclaim their clinical autonomy and learn again to trust their own judgments and the experiences of their patients. It is not acceptable to tell a patient in the midst of a drug reaction that the drug could not have caused his or her symptoms because those side-effects are not listed in the medical literature. Likewise, it is unacceptable to tell patients seeking safer and more effective treatments that there are no other options simply because EBM does not recognize holistic medical alternatives. Some of the greatest medical discoveries were made by independent-minded pioneers who were not afraid to buck the status quo, well before the dawn of EBM.

First-hand experience must be returned to its rightful place at the head of the evidentiary pyramid. When at odds, the course of action demanded by individual medical circumstance should always trump the generalized recommendations of RCTs. Progress, or the lack thereof, as reported by patients themselves should be elevated to the top of the hierarchy of empirical evidence. It follows then, that individual case studies reported by physicians should come in a close second, right behind patient's assessments of their own progress.

Experience is not anecdotal; it is primary. Rational abstraction is secondary, and is just as unreliable and subject to distortion, if not more so. Subjective information does not have to be relegated to the status of second-class citizenship. When combined with open-minded inquiry, ethical principles, and sound judgment to guard against inappropriate bias, the personal experiences of patients can provide valuable input that complements quantitative data and the observations of clinicians, thereby leading to more successful patient outcomes.

13. Successful short and long-term outcomes in medicine are more important than the certainty of its claims or reality of its left-brain oriented worldview.

Although science thinks of itself as unbiased and denies that it is interested in absolute certainty or ultimate truth, it contradicts those claims when it rejects a variety of alternative disciplines and theories on the grounds that they are not scientific. Because of its ability to measure and quantify, science likes to think that its methods alone are qualified to ascertain fact. Other methods are not only inferior; they are not science at all. But the practical reality of science is far from the theoretical ideal that it presumes to uphold.

Science was conceived as a method to investigate the world around us, not as a means of attaining certainty. The need to be certain is a prejudice that modern man imposes upon science. It is a function of left-brain imperialism, which desires to split the world, in accordance with dualistic theory, into black and white, right and wrong, and certain and uncertain. Having achieved that, it then dispenses with the subjective half of existence in the name of objective fact-finding. It defines its rules of evidence according to the various *–isms*, all of which are assumed to be true even though there is not a speck of scientific evidence to support the correctness of those presuppositions.

Neither is it the role of science to determine the ultimate nature of reality. Science's belief in its ability to reveal truth stems from a growing arrogance and mistaken belief that it has discredited the role of and need for religious influences in society. The voices of contempt for non-scientific versions of truth grow ever louder in the public arena. As a result, science itself is dangerously close to becoming a form of religious belief. And it is a role that science is uniquely unqualified to fulfill. As science's confidence grows, so does its intolerance for other legitimate sources of knowledge. I am not claiming that all or even a majority of scientists are of this ilk, but the overall collective voice of science is trending in this direction—and it is a very disturbing trend. Science would do well to develop a genuine sense of humility and respect for other methods, perspectives, and sources of knowledge.

Medicine follows suit in claiming possession of medical truth when it disparages other, so-called pseudoscientific health disciplines. When its monopoly on factual truth and certainty are challenged by unconventional healing disciplines, medical science defends itself on the grounds that—wait for it—only it is scientific while all others are not. This predictably circular argument wears thin rather quickly.

Medicine fails to comprehend that its mission is neither to define the nature of reality nor to know things with mathematical certainty. Certainty is not tantamount to science. Evidence is made no more or less scientific by virtue of its certainty or uncertainty. Science is supposed to be about discovering the world as it is, without imposing one's values on it. Its job is to shed light. It is not about defining the world in materialist, mechanist, reductionist, objectivist, left-brain, quantitative terms. And the imposition of those values does not in any way guarantee the truth or certainty of claims made.

Orthodox medicine does not have a clue when it comes to unconventional medical perspectives. It does, however, provide a particular perspective regarding its claims—a conventional medical perspective—which turns out to be just one perspective among many. Medical science will be fated to remain in the dark until the day it recognizes the limitations of perspective that come with its various –isms.

So, if medical science is not about truth, certainty, or reality, then what is its role? Its purpose, of course, should be the practical restoration of health and well-being to the sick *by whatever means possible*. Its role is not to be the medical police, enforcing its version of medical reality on patients and practitioners. It should not deny other methods of healing their right to exist and to be used by those who willingly choose to do so.

Medical science should be about choosing the best available methods from all disciplines for the betterment of patients. Medical education should not be exclusive, provincial, or territorial; it should be broad and inclusive. Medicine should not hold the yardstick of certainty above the heads of other, less quantitative or less tangible approaches to healing. It should not apply an either/or, right or wrong litmus test to theories of health and illness. It will eventually have to acknowledge that the mystery of human illness is far from the hard science that it wishes it to be.

Medicine imposes its will on patient care in order to maintain the scientific illusion that justifies its short-term outlook on the treatment of disease. A medical discipline that defines itself solely in material terms is less capable of comprehending the bigger picture and the longer-term implications of healthcare interventions. Only when it relinquishes its grip on the certainty of its *–isms* will it become possible to entertain new possibilities.

Holistic medicine is not bent on eliminating conventional medicine—that would be a disaster. However, it is capable of vastly expanding the horizons of medicine in ways that would be highly beneficial to practitioners and patients alike. And although truth, reality, and right and wrong are issues best left to philosophers, ethicists, and metaphysicians, medical science would be well served by a serious and thorough examination of the non-scientific presuppositions that underpin its scientific beliefs and practices.

14. Medical science is not an ideology; it is a method of open-minded inquiry. It is crucial to understand the differences between science and scientism.

Scientism is the antithesis to science. Skeptics are fundamentalist ideologues who are determined to protect science from invasion by pseudo-scientific influences. In actuality, they are defenders of scientism. Both popular culture and mainstream science are increasingly infected by scientistic dogma that passes for legitimate science. The future of science itself will depend upon a clear understanding of the differences between science and scientism.

The issue becomes even more complex when we realize that conventional science is also subject to a great deal of bias as a consequence of its adherence to a long list of –isms. Medicine is no different because it, too, presupposes the same –isms and is susceptible to a wide spectrum of scientistic intolerance.

The best that we can expect for mainstream medicine is that it is practiced with conscious awareness of its flaws and limits. There are many holistically-oriented practitioners and patients who understand these limits and seek to move beyond them. A thoughtful conventional doctor, therefore, offers his or her services and recommends that patients explore alternative options when conventional methods deliver less than satisfactory results. This is one simple way in which the two medical worlds can work together.

Unfortunately, some doctors actively discourage patients from seeking unorthodox means of healing. They do this not because they are avowed skeptics determined to eradicate so-called pseudoscience, but simply because they are applying what they have learned, or not learned, from their medical training. Many are uneducated about holism, a good number are afraid to deviate from conventional standards of care, a few are just not that inquisitive, and some exhibit bias to varying degrees for a variety of reasons. Most faithfully follow the –isms of medicine but are unaware of organized skepticism's absolutist anti-alternative agenda.

Then there are those who consciously align themselves with skeptics. Even then, they do so with varying degrees of commitment. Some genuinely believe that skeptics are defending mainstream science and medicine, while others are well aware of the nature of scientism and pledge their allegiance to this extremist faction. The worst offenders exhibit knee-jerk responses to the mere mention of holistic medicine; they are the least informed and most willfully ignorant.

The same scientistic spectrum of unconscious bias and conscious awareness is evident among the general population. The same is true across the board in the media, from popular entertainment to opinion to news. Scientism inflicts the most harm when it finds its way into serious journalism. The only truly effective remedy is an educated populace that is wise to the tricks of scientism masquerading as science. But that is not an easy task; it is one of the purposes of this book. Even some of the most educated persons do not understand the differences between science and scientism. This occurs mainly because we live in the age of science, a historical epoch during which few stop to question the motives or integrity of those who claim the mantle of science.

So how is it possible to discern the difference between science and scientism? First and foremost, it is crucial to remember that science is not an ideology. It is a methodology designed to help us learn as much as we can about the natural world. The minute we hear of some person or group waging a campaign in the name of science against an idea, a body of knowledge, or some person who represents unusual ideas, we must be on the alert for foul play.

The hallmark of scientism is its need to discredit opposing viewpoints. Real science is tolerant of competing ideas and welcomes alternative hypotheses and inventive new theories. This does not mean that science accepts all things as fact. It does not, on the other hand, purposely go out of its way to disprove something that challenges its current level of understanding. It remains open to new possibilities.

When the scientific community gets into a tizzy over some doctor who proposes a new theory to account for the rise in cases of autism, for example, and then seeks to censure or ostracize that person, we can be sure that scientism is at work. Medical science is not an unassailable body of knowledge, impervious to all attempts to alter it or add to it. A correct scientific attitude is one of curiosity, of open inquiry, always on the lookout for ways to achieve greater understanding.

The second hallmark of scientism, therefore, is its claim to certainty. Scientistic absolutism brooks no opposition or dissent. It believes that it knows what it knows because it has been proven by science. Skeptics wave the flag of science with fundamentalist fervor as if science is some monolithic and impregnable body of evidence that protects them from the insecurities of a hostile and ever-changing universe. One gets the same feeling when encountering a skeptic as when facing a religious fanatic. To a skeptic, the purpose of discussion is not to understand differing perspectives but to convince others of the wrongness of their ideas. The certainty of scientism always leaves polarized factions in its wake.

This, of course, is in direct contradiction to holistic reality, and to the lessons learned from everyday experience. Life is uncertain; it is filled with ambiguity. While some hard sciences enjoy a significant degree of concrete certainty, human medicine is far from an open and shut case. Human health is unpredictable. The disconnect between medical expectation and patient reality is the main reason why people angrily blame the doctor when they get sick, or do not get well, or when a loved one dies. Even though it anchors itself to the ostensibly solid ground of materialist philosophy, medicine is constantly changing. And it becomes even more unpredictable when it acknowledges intangible factors like emotion, mind, spirit, energy, and consciousness.

This is precisely what skeptics fear: the loss of their ability to know the so-called facts with absolute certainty. And so they resist with all their might, terrified at the prospect that they may be wrong, that their ideology, science, might fail them. But medical science is not an ideology that provides certainty of knowledge. The profession would fare much better if it dispensed with the need to project an image of certainty. It would enable it to open its collective mind to new possibilities. The acceptance of ambiguity as normal and natural would free medicine from dogma and allow it to explore new frontiers.

Holism long ago embraced the unknown by acknowledging the role that consciousness plays in health and healing. As a consequence, it has accumulated a great deal of valuable knowledge that remains taboo within the medical mainstream. Holistic medicine continues to explore the dynamics of healing, getting glimpses of insight here and there, as it wrestles with the mysteries of psyche and soma, unhindered by the egoistic burden of needing to pretend to know for sure. Isn't that the way all authentic approaches to knowledge should be? Should we expect any less from an authentic approach to medicine and healing?

15. Authentic green medical healing transcends the *–isms* of conventional science.

Conventional medical science flies blindly, vaguely conscious of the *–isms* that compel it to pursue its haphazard flight plan, unable to provide any unified or coherent theory as to why it does what it does. When pressed for explanations, it falls back on the predictably circular argument that the course of its activities is guided by science itself. Convinced that medicine is almost exclusively if not purely a science, the medical profession fails to examine its own belief system, thereby guaranteeing its fate. It repeats the same mistakes, unable to acknowledge that they are mistakes or to understand why they are mistakes.

Authentic green medicine is far more than a science. It is an alchemical synthesis of science, art, craft, philosophy, myth, metaphysics, spirituality, and more, wrought from the fiery crucible of the trials of human experience. It would be lost if it were not fully aware of its own assumptions regarding the nature of life, health, illness, and healing. The great medical thinkers of times gone by would be perplexed at modern medical science's dissociation from values, metaphysics, and experience. They knew that it is not possible to separate psyche from soma. This book was written to shine a light on the deficiencies in medical thinking that prevent us from developing a truly effective system of health and healing.

The future course of medicine will transcend the unconscious *–isms* that stand in the way of its maximum potential to heal the sick. This begins with the recognition of science's fundamental misperception of life as a material phenomenon. Medicine must give more than lip service to the reality of consciousness; it must incorporate it into its principles and practice.

Medicine flies blindly when its reductionist bias causes it to focus on body parts without taking into account their relationships to the whole person. It is not enough to go halfway, practicing a superficial form of holism by tending to the *whole physical* body. Genuine holism is green; it encompasses body, heart, mind, soul, family, community, and ecosystem.

Medical mechanism is compatible with a material worldview. It is easy to believe in cause and effect explanations when we can see and touch them. But like atomic particles careening around in space, material mechanism is an outdated construct. By limiting itself to such a simplistic understanding of disease etiology and treatment, medicine fails to take advantage of vast fields of knowledge that could revolutionize its approach to the sick and suffering. It fails to grasp that most human illness—most but not all—has its origins in the human psyche, and the so-called mechanism by which this takes place is, by its very nature, intangible.

Medicine makes a costly mistake when it discards experience in favor of the purely rational. In doing so, it falls prey to the stereotype of the egg-headed scientist who lacks all semblance of common sense. The machinations of the logical mind are just as capable of misleading us as are the subjective impressions of an ungrounded person. A balanced and judicious mix of reason and experience, however, provides a more effective path to knowledge. Together they increase the odds of acquiring a thorough holistic comprehension of the true nature of health and healing.

Reliance on objective information alone is a self-imposed handicap that medicine must eventually relinquish. The inescapable irony is that is not possible to be objective while excluding all that is subjective. It is also an illusion to believe that health and illness are objective "things." Healing is a process that takes place within a broad spectrum of subjective experience. To objectify illness is to sever it from its meaning, from consciousness, from life itself.

The contemporary definition of medical empiricism is a twisted knot of logic that serves to reinforce medicine's objectivist perspective. To make observations regarding the material dimension of illness while leaving out subjective impressions of the immaterial is to paint a two-dimensional portrait in black and white. To bring it to life, one must add the color and depth of personal experience. It turns out that it is quite possible to make subjective observations of intangible phenomena. Authentic medical empiricism entails acknowledgment of all contents of consciousness, not just observations of the physical phenomena recognized by materialist science.

Belief in the one correct way of solving a medical problem is a canard borne of medical egotism. Medicine's territorial imperative is unbecoming of true science and a roadblock to innovation and progress. No academic, pharmaceutical, biotechnological, insurance, professional, or corporate entity should have exclusive rights of ownership over the healing profession. Diversity is a universal characteristic. Humanity has developed a multitude of languages to communicate, numerous methods and styles of artistic expression, and a colorful variety of ways of expressing awe at the transcendent nature of creation. Amidst that diversity, in each case there is a common purpose that unifies—communication, expression, reverence, and wonderment. Medicine and healing are no different. The mysteries of human illness are best handled with a diverse bag of methodological tricks, all of which have one common aim. That aim is healing. The real purpose of healing is the growth and maturation of the human psyche as it becomes increasingly conscious through the trials of illness.

Although duality is a very common experience, so, too, is unity consciousness. Division is an unavoidable consequence of the rational mind, which is incapable of grasping the whole in its entirety without some assistance. Subject and object, body and mind, and healer and healed are but surface aspects of a deeper unified reality. That reality becomes accessible only when we suspend the turnings of the rational mind in order to enter into the quiet space beneath. When we direct our healing efforts to exterior manifestations of illness, the outcomes are often temporary and superficial. Healing aimed at the interior, directed toward deeper causes, to the extent that they can be discerned, yields more lasting and meaningful results.

Authentic healing involves much more than just the rational faculty. The knowledge provided by evidence-based medicine is inadequate. It represents the hypertrophy of the rational function to an extreme, to the exclusion of all other faculties of human experiential knowledge and wisdom. The capacity for deeper healing increases to the extent that we assimilate knowledge derived from the wider spectrum of non-rational direct experience.

Science is not an ideology; it is a methodology, a tool for acquiring knowledge. Science and spirituality are mutually exclusive only in the minds of dualists and materialists. It is possible to be a good scientist while simultaneously dedicating oneself to the spiritual life. A scientific career can even be one's spiritual calling. Religious fundamentalists mistrust science because they fear it is out to discredit their beliefs. They are rightly concerned, however, the real threat to religious freedom originates not with science, but with scientism. Far left skeptics are intolerant of religion because they have unconsciously adopted science as their religion. Right-wingers' questioning of science is really a reaction to scientism. Real science that understands its limits poses no threat to religion—and genuine spirituality should have no problem with science that is aware of its boundaries.

Authentic science is not hung up on certainty. Certainty is an ideological value imposed upon science by modern man. The need for certainty grows in importance to the extent that we reject religion or spirituality. It compensates for the insecurity that comes from having no belief system to confer meaning or purpose to human existence.

Neither is science about universal truth. It offers only one particular construction of reality. There are numerous other, equally valid constructions of so-called reality. Science fails when it attempts to be an ideology, an arbiter of truth. This is a critically important point because health, illness, and healing have everything to do with beliefs, values, awareness, consciousness, and spiritual growth. In denying this, medicine leaves us with nothing but cold, hard, meaningless facts.

Authentic green science transcends the boundaries of conventional science, not in the sense that it knows better, or is more certain, but in the sense that it is willing to explore previously forbidden frontiers. It does not flinch at the implications of psyche, spirit, or consciousness because it understands that body and mind are one.

Conventional medicine must respect the individual personal truths of its patients, confine itself to discovering practical and effective ways of healing the sick, and allow for the unobstructed emergence of a new paradigm of healing that incorporates physical medicine, subjective experience, and all aspects of psyche and spirit.

CHAPTER 16

A Clinical Case Study

It is necessary to know sickness, not from pathology, not from physical diagnosis, no matter how important these branches are, but by symptoms, the language of nature.
–J.T. Kent, MD, *Lectures on Homeopathic Philosophy*

The following case history is offered as an illustration of the real-world impact that our presuppositions can have on the practice of medicine. This particular case involves a homeopathic methodology, which is then compared and contrasted to the standard medical approach. Our *a priori* non-scientific metaphysical assumptions regarding the nature of mind, body, knowledge, and existence play crucial roles in the type of medicine we choose to support and the kind of treatment we come to expect. Authentic green medical science is not bound by ideological *–isms* that prevent it from entertaining new ideas, connecting the dots, and trying new approaches to health and healing.

Case History

A patient who I had originally seen for a few visits three years earlier returned to my office for assistance with a new health problem. I had previously treated Marie, a 56 year-old married female nurse practitioner, for nausea, vertigo, and constipation. A series of homeopathic prescriptions had helped resolve most of those difficulties.

Now Marie presented with a chief complaint of sciatica, which had been bothering her for months. She complained of pain originating from the left sacroiliac region and extending down the left leg all the way to the foot. Questioning revealed that the pain was made worse from sitting and better by walking. Standing aggravated the pain to a lesser extent and the application of an ice pack helped alleviate the problem to some degree. She also noted that the muscles of the afflicted leg occasionally twitched. Diagnostic imaging had revealed nothing unusual.

Per the advice of her regular doctor, she tried Advil, which provided only temporary relief. Although it was recommended, she never got around to doing physical therapy and she refused the one other option offered by her doctor, prednisone. Marie's nursing background had made her aware of the potential dangers of this powerful immunosuppressive drug.

When asked about factors that may have led to the onset of her sciatica, Marie noted that she had run in a 5K race one week earlier. She also talked about the stress resulting from working long hours at her new job over the past four months. I asked her how the stress was making her feel. She replied, "On work days I wake with the feeling that something terrible will happen. I feel personally responsible for everything. It's tied to my daughter's death seven years ago."

I looked back at the original full intake that I had conducted three years ago. She had reported a history of an episode of shingles that she attributed to the stress of losing her daughter. It consisted of "painful lesions" on her left little toe extending to the front of her lower left leg, with pain shooting up the leg. I wondered about the potential connection to her present complaint of left-sided sciatica.

Assessment

A homeopathic evaluation often places particular emphasis on the more prominent symptoms and any modifying factors that affect those symptoms. In Marie's case, it was important to know the *modalities* that aggravated and ameliorated her sciatic pain. Once this was ascertained, it became possible to consult a symptom index called a *repertory* in order to determine what substances were known to be able to cause, and therefore to treat, those symptoms. This illustrates the essence of homeopathic treatment, which employs small doses of substances capable of mimicking patients' existing symptom patterns. The closer the match between the medicinal symptom profile and the patient's symptom profile, the more satisfactory the results tend to be. Based on our discussion, I chose to look up the following symptoms in my homeopathic repertory:

> *Left-sided sciatica*
> *Sciatica aggravated from sitting*
> *Sciatica ameliorated from walking*
> *Sciatica aggravated from standing*

Two different homeopathic medicines keyed out strongly, each known for its ability to produce, and therefore treat, all four symptoms. One was *Kali bichromicum*, which is the Latin name for a medicinal preparation made from potassium bichromate. The other was *Kali iodatum*, a preparation made from potassium iodide.

With this in mind, I asked a few more questions in order to further narrow my choice down to the medicine that I thought would be most appropriate for this particular case of sciatica. Knowing that iodine-containing homeopathic medicines are often indicated in the treatment of thyroid conditions, I made a mental note of Marie's long, thin body type and asked her if she had a high or low metabolism. "High, I can eat anything and never gain weight," came the reply. I asked but she denied ever having any thyroid-related health problems. Given the repertorization results and Marie's high metabolism, I decided to prescribe *Kali iodatum* 30c (30c is a measure of the dilution factor of the medicine in solvent), two doses only, once in the morning and once in the evening.

Follow up

Marie returned to my office three weeks later, "It was very interesting. I feel better. I feel more energetic. I went back to work (that day after visiting my office) and had the worst pain yet. It was like lightning in my leg, but it only lasted for 45 seconds."

She experienced a similar but less intense sensation several times during the following day for shorter durations. After that, the sciatic pain completely disappeared, with the exception of two brief episodes. One occurred when Marie's boss pointed out a mistake that she had made at work. The second episode was triggered when a co-worker, who was aware of Marie's general unhappiness over her work situation, suggested that maybe Marie wanted to try working on a part-time basis.

This led to an illuminating discussion about the circumstances of Marie's employment. It turns out that she had been working for a medical facility designed to help patients overcome their alcohol and substance addictions. She originally sought employment there as a way of turning the negative of her daughter's drug-related death into a positive experience. In fact, Marie wanted her daughter to seek help at this same facility years earlier but, to Marie's great grief and disappointment, she had passed before she got a chance to do so.

Marie's six months of employment at the substance abuse rehabilitation center had the unintended consequence of reminding her on a daily basis of the tragic circumstances of her daughter's death. "I hate that job. I gave notice." Soon after the sciatic pain had lifted, she came to the conclusion that the job was no longer worth it. It was not fulfilling her needs and it was becoming an increasingly painful place to work. The day after she gave her boss notice of her intent to leave she became "achy, feverish, felt freezing cold, and went to bed with a hot water bottle." She recovered within 24 hours, returned to work, and told her boss of her desire to inform her fellow co-workers about the reasons for her unexpected resignation from the job.

The boss thought otherwise but Marie insisted. A meeting was arranged and Marie stood in front of the rehab center's staff. She told them for the first time the story of her daughter's premature death and of her hope that working there would give meaning to both her and her daughter's life. But it had become too difficult and that was why she was leaving. "I told them the truth. It was hard to hear similar stories from other parents who had lost their children to addiction." She was greeted with sympathy and felt relieved at the decision she had made.

It was clear that, in addition to relief from her sciatica, Marie was undergoing profound changes in her psychological perspective. I felt that it was premature to make any therapeutic changes. Her situation was in flux and improvement was taking place on all levels, mentally, emotionally, and physically. I explained the need to be patient, that we needed to let events unfold, and gave her two more doses of *Kali iodatum* 30c to take home with her as a precaution. Over the next few months, the same medicine was taken a couple times more. Marie continued to do well, her sciatic pain cleared up, and she moved on to a new job.

Discussion

In my estimation, the near instantaneous resolution of Marie's months-long bout of debilitating sciatic pain had occurred while simultaneously having an epiphany about the true source of her pain, which she attributed to the burden of keeping the secret of her daughter's drug-related death from close associates in the workplace. Once she unburdened herself of the secret and resolved to leave the job, the pain vanished. Regardless of whether you believe there was a cause-and-effect relationship here, this *is* what happened. It would be decidedly unscientific to overlook this sequence of events and its potential ramifications.

Now the usual skeptic's response would be to dismiss this case history as anecdotal and claim that Marie's improvement was purely spontaneous, having nothing to do with the homeopathic medicine or Marie's job situation. That would typically be followed by a demand for scientific proof as to the effectiveness of the chosen homeopathic medicine, conveniently overlooking the fact that the case itself is a valid form of clinical evidence. It would be easy to dismiss the action of Marie's homeopathic prescription as a placebo response, but that would have to be done in disregard for the fact that, over the past two hundred years, homeopaths have observed the resolution of thousands of similar cases of sciatica after prescribing *Kali iodatum*. It is the reason why this particular medicine is listed in homeopathic reference books as an indicated treatment for a specific pattern of sciatic symptoms.

Let us dispense right away, therefore, with all disingenuous knee-jerk reactions that have little to do with science and everything to do with a desire to discredit anything that falls outside the lines of mainstream medical understanding. Instead, we document Marie's symptoms in a non-judgmental way and take her reports at face value. Realizing that an overly skeptical attitude would prevent us from learning anything new, we then proceed in an open-minded manner to ask questions about the nature of the phenomena observed in Marie's case. This would be more along the lines of my understanding of genuine objectivity.

Was Marie's sciatica a function of physical factors or was it due to psychic distress? Far from the exception, such events are commonplace in the experiences of most holistic practitioners. The practices of conventional physicians are filled with similar cases but their philosophical orientation precludes them from viewing such situations in other than materialistic terms. Is it appropriate to dismiss events like this, writing them off to coincidence, thus severing mind from body, because we cannot provide tangible scientific proof of their connection, because such phenomena fall into that black hole, the explanatory gap that continues to confound materialist medical theory? Or is it time to make room for the role of consciousness in our quest for knowledge regarding human health and well-being?

Even if we were to dismiss the impact of the homeopathic prescription, we would still have to account for the psychosomatic nature of Marie's sciatica. Marie, herself, understood the connection between her daughter's death and the sciatic pain, but apparently needed a little help, or a push from a homeopathic medicine if you will, before she was able to overcome the psychosomatic inertia that kept her from taking decisive action to resolve the problem. The pain had served its purpose to nag Marie enough to prompt her to do something about her situation. Once it clicked in her mind, prompting her to take action, the pain was no longer necessary.

A materialist approach to this case would involve a search for physical causes and the implementation of physical solutions. Imaging technologies would be employed to search for bulging discs, tumors, and/or arthritic deformities. If this failed to reveal an overt concrete cause, it could always be attributed to a "pinched nerve," one of those satisfyingly graphic throwaway explanations that serve to keep the focus on the material plane. Naturally, Marie's psychological state and job situation would be meaningless in this context.

From a reductionist perspective, a pinched nerve, bulging disc, or vertebral deformity would serve the need to place the blame on a specific, localized part of the body—in this case, an anatomical part, in other cases, a biochemical pathway or physiological process. A broader explanation involving psychological states and life circumstances would be far too general to satisfy a reductionist philosophy of disease. This is not to say that there are not plenty of people out there who suffer from pain resulting from physical causes—it's just that, in Marie's case, physical causation was last on my list of suspicions.

The same philosophical conundrum would apply to mechanistic causation in this case. There are no mechanisms, physical or otherwise, that account for or explain how or why Marie's sciatic pain could be connected to or triggered by her states of mind. And yet, here again, the mind-body problem demands to be acknowledged. If an answer in material, reductionist, and/or mechanical terms is not forthcoming, this is an indication that we need to reevaluate our philosophy of medicine. Experience tells us there is no real mind-body problem as much as there is a philosophical misconception regarding the nature of mind-body. There is also a great deal of psychological resistance to accepting the realities of mind-body dynamics.

The pinched nerve theory is a perfect example of the rationalist influence in medicine. In the absence of any evidence of physical causation, a logical and materially plausible explanation is formulated in order to satisfy the needs of both doctor and patient to frame the affliction in concrete terms. Medicine is riddled with rational assertions of this nature. They fulfill the rationalist need for logically tenable explanations, regardless of their scientific veracity. Again, I am not saying that it is not possible to have a pinched nerve, but I am saying that in the majority of instances there is no actual proof to back up such claims. Nevertheless, it passes for sound science simply because it seems to make logical sense.

From an objectivist perspective, Marie's job circumstances and emotional state would be, by definition, automatically excluded from scientific consideration due to their unreliably subjective nature. It is only possible to trust the objectivity of our scientific instruments and the doctor's first-hand observations of physical evidence. We respect patient's reports but take them with a grain of salt because we are supposed to believe that patients cannot be as objective as scientists. As per the objectivity criterion, one would think that pain itself *should* be excluded from investigation because, apart from its neurological correlates, all other aspects of pain are purely subjective. But pain, the most ubiquitous patient complaint of all, is apparently important enough to be exempted from the usual standards of scientific rigor—otherwise, most medical ailments would be judged to be mere figments of patients' imaginations.

Empirically speaking, given the absence of radiological findings, there is no observable concrete evidence other than the facial grimaces and cries of discomfort that could be elicited by manipulating Marie's leg in certain ways in order to reproduce her sciatic pain. Remember that the contemporary definition of empirical evidence does not include observations of subjective phenomena made by the physician or patient. Homeopathic practice is predominantly, but not exclusively, an empirical methodology. As with the Empiric physicians of old, this type of experience-based medicine does not meet modern standards of medical empiricism. Homeopathy, therefore, along with a host of other empirical holistic approaches, are usually dismissed by medicine on ideological grounds, regardless of the outcomes that they produce.

As an aside, it should be noted that most holistic practitioners would not view Marie's flu-like symptoms as a coincidence. When a person experiences profound psychic and/or physical changes, the innate heal-ing intelligence of the mind-body will often generate what is commonly known as a healing crisis. Symptomatic flare-ups in this context can be understood as manifestations of the life force's self-healing mecha-nism. A person who experiences a healing crisis will often emerge from an acute symptomatic flare-up with a new perspective, not unlike some-one who claims to be born again after having endured a trying hardship. The same healing crisis can signal the beginning of the end of chronic health problems.

In Marie's case, the sciatic pain persisted until she gained new aware-ness of the nature of her struggles, at which point the life force assisted the changeover by redirecting her chronic sciatic symptoms into an acute but temporary episode of flu-like symptoms. The initial lightning-like aggravation of sciatic pain can be understood in the same way—it peaked before it began to resolve by shifting in the direction of fever and chills.

This is the sort of phenomenon that conventional physicians fail to comprehend, primarily because they do not have a holistic frame-work that enables them to connect the dots between successive medical events. It also provides evidence that challenges the dualist perspective of mind and body as separate entities. This case vividly illustrates how mind and body operate in unison as one seamless whole.

With all that said, it must be made clear that the homeopathic methodology used in Marie's case relied almost entirely on subjective information. Her own descriptions of the pain, its location, when it was aggravated, and when it was ameliorated, along with the situational and psychological context of her affliction, constituted the evidence upon which treatment was based. It was, in essence, the information de-scribed by the patient herself. The same type of subjective information also helped determine whether the treatment was a success or not.

This is just one example of a viable alternative scientific methodology that is capable of making sense of subjective information and using it for practical purposes. Empirical observation is employed to gather both objective and subjective information, which is then put to the practical test in a trial and error process of treating patients. Over the decades that process has become increasingly refined, documented and built upon the prior clinical experiences of thousands of practitioners. Theories regarding the nature of illness, healing, suppression, palliation, and susceptibility and resistance to illness are constructed from the practical lessons learned from treating many patients.

Note that, in this case, theory follows experience, not the other way around. Evidence is not gathered for the purpose of verifying a theory that has been formulated beforehand from speculative logic. Empirical evidence obtained from the outcomes of treating patients provides the foundation for the formulation of theories that seek to make sense of the evidence. It should not be necessary to produce evidence to support a theory that has no prior basis in fact; theories are proposed after the fact to support the existing evidence.

This is just one scientific method among many. Homeopathic methodology does not shy away from subjective information, which is believed to be complementary to objective data. Together they form a more accurate picture of holistic reality. From this perspective, science is not a monolithic ideology that dictates allegiance to one particular methodology; it is a tool of inquiry that employs many methods.

Please also note, that had Marie not already had a conventional evaluation to rule out potential physical causes, I would have recommended a workup, either initially or in the event that my chosen treatment had failed to bring relief. It is quite reasonable, I believe, depending upon circumstances, to initiate treatment first and then resort to expensive and time-consuming diagnostics later, when and if that treatment yields unsatisfactory results.

There are many such ailments that refuse to bend to materialist standards of physical causation. I have treated cases of asthma that were triggered by psychological factors and migraine headaches that had their onset with emotional events. Cases involving muscle spasms and back pain are notoriously psychosomatic. Even flu-like symptoms can be triggered by psychologically trying experiences. There are many variations and very few rules, contrary to what the conventional taxonomy of diagnostic categories and their criteria would lead us to believe.

It is necessary to cut through ideology in order to perceive patient reality. When physicians acknowledge the importance of that reality in their medical calculations, the likelihood of therapeutic success increases accordingly. An authentic approach to health and healing entails an open-minded attitude that is willing to suspend existing beliefs in order to entertain innovative ideas that normally go unrecognized and that can contribute to more satisfying patient outcomes.

CHAPTER 17

The Rebirth of Medical Philosophy

Man's exploration of nature and its processes gave rise to the theory of what we now call science. In olden times, these researches were not carried on by scoffing materialists, but by enlightened philosophers who discovered, beneath the superficial aspects of form, divine chemical and mechanical processes. Science is therefore the study of the anatomy of the body of God.

–Manly P. Hall, *Words to the Wise*

The whole point of science is that most of it is uncertain. That's why science is exciting—because we don't know. Science is all about things we don't understand. The public, of course, imagines science is just a set of facts. But it's not. Science is a process of exploring, which is always partial. We explore, and we find out things that we understand. We find out things we thought we understood were wrong. That's how it makes progress.

–Physicist, Freeman Dyson

Science and philosophy must flourish together, for it is not possible to be deeply informed in one without an equal understanding of the other.

–Manly P. Hall, *Words to the Wise*

M odern science is not always what it claims to be. It is increasingly faith-based, it is not conducive to freethinking, and medicine in particular is hampered by this troubling trend. The rise of science in the West as the arbiter of truth roughly parallels the decline of philosophy. The capacity to think deeply on subjects of great weight has been gradually replaced by an unreflective faith in the facts brought to us by science. Some argue that the presumed certainty of scientific knowledge has rendered the need for philosophical inquiry obsolete. But it is philosophy that makes the attainment of wisdom possible, an achievement that scientific fact alone is incapable of providing.

Western medicine eagerly embraces objective information related to human health but its materialistic bias prevents it from recognizing less tangible factors regarding energy, spirit, meaning, and consciousness to advance the well-being of patients. Dreams, synchronicities, spiritual insights, psychosomatic and psychic phenomena, and even physical symptoms that do not make mechanistic sense, therefore, are not recognized as legitimate research topics and are not factored into patients' medical evaluations. Medicine's reductionist bias favors a fragmented system of medical specialties that tends to the physical ailments of isolated parts of the human body, while it prejudges a multiplicity of holistic approaches as untenable, unscientific, and unrealistic. It welcomes mechanistic explanations compatible with its cause-and-effect perspective, but it doesn't have the foggiest notion of what to do with a multitude of phenomena that cannot be explained in conventional medical terms.

The rational bias of Western medicine predisposes it to devalue qualitative phenomena, such as the subjective experiences of patients, in favor of quantitative data that can be measured and manipulated. Empiricism, likewise, is defined in such a way as to exclude the subjective experiential reality of sensory observations. Like science and medicine, Western philosophy is grounded in dualism and therefore susceptible to the same errors of thinking. It will require a somewhat radical departure from Western thought, therefore, to rescue science, medicine, and philosophy from their innate biases, from the *-isms* that confine them.

The common perception of philosophy as an idle exercise in navel-gazing is, in part, a consequence of science's imperialistic overreach into the realm of meaning, purpose, and values. Science has squeezed philosophy out of the picture, while philosophy has passively assimilated the values—if they can truly be called values—of scientific materialism. Lacking the courage to stand on its own, modern philosophy has assumed a posture that is largely in sympathy with the conventional scientific perspective.

Any truly authentic approach to healing must be cognizant of the presuppositions that inform its worldview. To do otherwise is akin to navigating the seas without a compass. Conventional medicine is convinced that questions regarding basic assumptions were settled long ago by science. Having dispensed with the need for self-examination, it set its course on cruise control, and is not particularly inclined to look back. The dearth of new ideas in medicine is a direct reflection of the moribund state of medical philosophy. Genuine innovation is no longer possible because underlying beliefs regarding the nature of health and illness remain static. Medicine will be unable to evolve until its metaphysical framework evolves first.

Lacking the necessary navigational compass, medicine continues to pile fact upon fact, unable to synthesize the lessons lying dormant in its storehouse of information into any coherent philosophy of health, illness, or cure. The result is a confusing and fragmented array of disjointed specialties, theories, and oftentimes hazardous practices. The only way to sustain the grand enterprise is through mental trickery, via an overly rationalized logic that continues to justify the benefits regardless of the risks. Gradually the goalposts are moved in favor of perceived benefits while downplaying the dangers. Lacking the capacity for self-evaluation, the medical behemoth rolls on, increasingly wedded to corporate interests that have little or nothing to do with patient interests.

Modern academic philosophy has achieved a reputation for irrelevance for good reason—it dissociates itself from values, from real life issues—and this is a function of its intellectual allegiance to the *-isms* of conventional science. Like science, it strives to be objective and detached, almost clinical in its approach to human concerns. Philosophy is perceived by many to be little more than an empty exercise in circuitous logic and there is a good bit of evidence to support this stereotype. It seems that philosophizing has devolved into the ability to compose clever retorts designed to poke holes in opponents' arguments—the use of intellectual tricks of logic that reward the those can contrive the most abstract contingencies to throw others off the trail. Caught up in the art of argument, philosophy seems to have forgotten the larger questions of life that give meaning and purpose to existence. Having distanced itself from values, modern philosophy has become an intellectual parlor game, a form of mental gymnastics for an elite class of academics. For the common person, philosophy is all but dead. Of course, this is only partially true. There are always pockets of resistance, but one has to go out of one's way to find them.

Although the profession may believe that the fundamental principles of medicine are set in stone, the healing of illness is far from a settled matter. The belief that medicine has a lock on how to best handle disease is a function of a great number of erroneous assumptions. Medicine is so confident of its abilities that it dismisses most new ideas not in sympathy with its worldview without giving them serious consideration. It does so because it presumes that there is no other perspective that can achieve successes comparable to those achieved by contemporary medical scientific methodology. In doing so, it violates scientific principles of open-minded inquiry and discovery.

In making such assumptions, medicine creates an atmosphere that squelches freedom of thought. There is no need for new ideas; it is not necessary to reinvent the wheel. All that is needed is to continue to fill in the knowledge gaps with the existing tools that medical science already provides. Anything else, whether they be energy healing techniques, indigenous healing traditions, or alternative medical systems such as acupuncture, homeopathy, or Ayurveda are believed to be inferior because they are not based in medical science—science, that is, as defined by a particularly rigid set of –*isms*.

While some alternative healing modalities may not be compatible with conventional medical science's methods, they do, nevertheless, represent accumulated bodies of knowledge and wisdom, which have their own coherent methods that make perfect sense to those trained in such disciplines. Whether it is acknowledged or not, there is no one single scientific method. There are a variety of methods. There are also non-scientific avenues to knowledge that can provide valuable insight into the human condition, health, and illness.

It is rather puzzling to consider the enormous body of medical information, data, and statistics, and then realize that there is no mainstream equivalent of a medical think tank. There is no truly relevant department of medical philosophy at any conventional medical school that works in a meaningful way with the medical profession. (The exception, here, is medical ethics, which I will discuss shortly.) How is it that we have arrived at this point without having given substantial thought to what it is that medicine does, why it does it, and whether it should be done? Are we to believe that science renders the need for careful thought and self-reflection moot? Science seems to have adopted the misguided position that it can't allow unscientific ideas to interfere with scientific facts. After all, it is reasoned, philosophical ideas are untrustworthy because of their subjective nature.

Nevertheless, medicine is far from a settled matter. This is clear when one considers the diversity of historical, cultural, allopathic, and alternative beliefs regarding health, illness, and how it should be approached. The status of conventional medicine is comparable to that of Newtonian physics, which is still relevant and has practical applications under certain circumstances, but has been eclipsed by a more advanced form of quantum physics that has broader applications and deeper implications. Both have their roles, but most of the action and the greatest potential for breakthroughs lie with quantum physics.

Most holistic medical modalities employ advanced energetic concepts that material medicine is not capable of grasping given the conceptual restrictions of its *-isms*. Holism does not replace the need for Western medicine but it does render many conventional practices unnecessary. It also significantly expands the potential for genuine, lasting healing of body, heart, mind, and soul.

If medicine is ever to live up to its true potential, it must first return to its roots to re-examine beliefs that were long ago presumed to rest upon unbendable scientific truths. What we need is a return to freethinking. Medical schools should not be centers for indoctrination that discourage enthusiastic students from exploring new frontiers. They should encourage the best and brightest to be creative thinkers. We cannot think creatively if we do not understand the manner in which our beliefs inform the practice of medicine. We must re-examine the assumptions that undergird our belief systems and be willing to consider new possibilities, new worldviews, and new realities. And this cannot be accomplished without the aid of medical philosophy.

We have reached a point where science has departed from its own standards, and where the general public accepts most things uttered by scientists as scientific truths—regardless of their basis in science. Science is gradually morphing into a form of scientistic religious belief precisely because freethinking is no longer tolerated by the scientific mainstream. Freedom of thought in science has been replaced by hero worship of scientists and technological innovators, not unlike the idolatry reserved for the gods of religions that forbid defiance of their authoritative edicts. This perception of science gone astray is no longer a minority position as many are becoming concerned about the lack of restraint represented by scientific imperialism. Here is biologist Rupert Sheldrake on the subject:

> *Much of the hypocrisy of science comes from assuming the mantle of absolute truth, which is a relic of the ethos of absolute religious and political power when mechanistic science was born. Of course, there are disagreements among scientists, and the sciences are continually changing and developing. But a monopoly of truth remains the ideal. Dissenting voices are heretical. Fair public debates are alien to the culture of the sciences.*[59]

The prerequisite suppositions that define sound science and medicine are really biases that limit the boundaries of exploration. Constrained by the conventional –*isms*, we are left with a superficial understanding of the material body decoupled from the realities of consciousness, the mind-body unity, and the dynamics of healing.

Science deals in profane fact, not transcendental principles. Our understanding of the higher truths can have an enormous impact upon the way in which we live our lives and the way that we approach our mental, emotional, and physical health. Medical science has nothing to say regarding higher truths. Its realm is the mundane. Conventional science is a science of the mundane that simultaneously presumes to bypass the need for contemplation of higher metaphysical principles—principles without which science would have no foundation. In light of these higher truths, it would be unwise to suspend one's critical judgment simply because science claims that the facts are settled and not subject to revision.

Our basic assumptions regarding life, death, reality, health, and illness make all the difference in the medical strategies that we employ. A renewed medical philosophy must return again to the fundamental questions regarding the nature, purpose, and methods of medicine. It must also clarify terms—such as healing, cure, disease, treatment, palliation, and suppression—whose meanings are often confused in the minds of both the general public and medical professionals. Plunging ahead in disregard for the answers will result in more of the same haphazard, symptom-oriented, knee-jerk, putting-out-of-fires approach to treatment that never seems to get to the bottom of things.

We can ask broad metaphysical questions that influence our overall orientation toward human illness. If there is more than this material existence, then how can the physical body be the only factor relevant to health? Does mind exist at all or is the physical brain the only reality? Are there individual minds, can there be collective psychic influences, and how does human consciousness relate to health? Are mind and body separate realities or is the mind-body one seamless entity? Is there an actual separate objective reality or is it merely a construct of the rational mind? Does subjective experience become irrelevant to health issues by virtue of the fact that it cannot be detected by the tools of science? Is the knowledge produced by science the only type of knowledge relevant to the treatment of illness?

We can also ask philosophical questions regarding the dynamics of health and healing from a more comprehensive, holistic perspective. Is a fever a desirable or undesirable thing? Is it a sign of health or illness, immunological competence or weakness? Is it a manifestation of self-healing? Is a fever merely random or is it a tactic employed by the body's bioenergetic defense mechanism?

Or we can ask philosophical questions of a more specific and local nature. Should this particular fever in this particular context in this particular patient be treated or left to run its course? Was a prior psychological stressor a contributing factor to the onset of this fever? Should a fever be treated simply because it is technically possible to do so? Can treating the fever undermine the self-healing intent of the body?

These larger metaphysical and philosophical questions regarding the nature of Western medicine and its inherent strengths, weaknesses, biases, and contradictions should not to be confused with medical ethics. On the whole, medical ethics tends to accept medicine as it is and does not question its fundamental premises. Ethics takes the conventional medical worldview and its –isms for granted.

We can ask ethical questions regarding the applications of medicine, as it presently is, without questioning the underlying scientific beliefs that support its practice. Is it ethical to prescribe stronger than recommended doses of a particular medicine, even though there is no evidence in the medical literature to support such an action? Is it ethical to offer kidney transplants only to those with adequate insurance coverage? Is it ethical to treat a fever simply for the sake of symptomatic relief, knowing that it may hinder the healing response, slow recovery, or lead to undesirable side-effects?

Medicine is content when philosophers focus their attention on ethics because it allows conventional medical beliefs to elude criticism. In this arrangement, it is not medicine itself that receives the scrutiny but the applications of medicine. Ethics is more narrowly focused. It concerns specific issues and instances and is thus prone to the biases of reductionist thinking. Since medical ethics is a specialized version of philosophy, and since its core beliefs are in agreement with those of conventional medicine, it does not pose a threat to the medical profession. Not surprisingly, medical ethics also assumes that conventional medicine is the only game in town. Other forms of medical treatment and healing are simply not on its radar screen.

Metaphysics is the broadest of philosophies. Its scope encompasses more than just the world of science or medicine. It is a philosophy that circumscribes matter, consciousness, and spirit. A metaphysical examination of core principles, however, is anathema to medicine, which is perfectly content to keep the balance of power just as it is. Questions regarding core principles can have profound implications; the kinds that have been known to lead to paradigm shifts in perspective and practice. Such questions, and their answers, are the kind that mainstream thinkers tend to resist.

Fortunately, medical philosophy is alive and well, mainly because it has found refuge in many alternative forms of medicine. Although holism is a methodological approach to healing, it is a philosophy first and foremost. It is a set of principles upon which the edifices of many healing modalities have been built. Holism as a value system, as a set of beliefs grounded in experience, precedes the facts of the sciences that it supports. The facts come later, and are dependent upon the lens through which one perceives circumstances. Metaphysical principles come first, scientific evidence is gathered later, and theories regarding the nature of health and healing are then proposed based on the available evidence.

The same applies to the facts according to conventional medicine; they are informed by the materially oriented values or –isms that conventional medicine places its faith in. In the end, it is not a question of whose version of the facts is truer. What matters most are the benefits that accrue to living, breathing patients.

The defining difference is that holism provides a complete perspective, a more thorough understanding of the whole than that represented by the isolated fragments of conventional medical theory and practice. There is philosophical relativity and then again there are metaphysical truths. A consistently effective system of healing cannot be built on relative principles that encompass only a part or parts of the whole picture. The broader the metaphysical foundation of a system of healing, the more closely it will be capable of approximating truth—as far as it is possible to ascertain truth—and the more successful and satisfying will be the results that it yields.

Medicine can and should involve more than just short-term repair of the physical body. The goal of all healing should be the thorough, lasting, and comprehensive restoration of health to each ailing individual, both in objective and subjective terms. At its best, healing entails personal growth, the raising of awareness, and the expansion of consciousness in ways that contribute to ongoing health of body, heart, mind, and soul. In service to that end, one cannot focus solely on the material while dismissing the role of values, meaning, purpose, energy, consciousness, spirit, and soul. More than mere scientific knowledge, all authentic forms of green medicine, if they are to be genuinely effective, also require a deeper understanding of the dynamics of healing.

Patients do not care whose version of reality is true, as long as they find relief from suffering. Truth is not the aim of holism, nor should it be the objective of conventional medical science either. It is not a question of whose version of reality is more accurate, but whether medicine is open-minded enough to adopt new methodologies that lead to better patient outcomes. Science as a whole is much better at producing practical results than it is at determining ultimate truth or the nature of reality.

Nevertheless, an important distinction must be made regarding practical ends. Material medicine tends to yield palliative, suppressive, and short-term benefits while holistic approaches are more capable of producing longer lasting results. They are complementary perspectives and each has its place under the appropriate circumstances.

It is also true that subjective patient satisfaction is an important component of healing. Commonly, the first sign of impending disease or illness is an individual's subjective sense of discomfort or disharmony. Regardless, it is not uncommon for a suffering patient to be told that there is "nothing wrong" because the objective medical workup reveals no abnormalities. Western medicine downplays the subjective experiences of patients because it views healing almost exclusively as a practical and technical matter. Similarly, the first indication of recovery is often a renewed sense of clarity, energy, and well-being, even before objective signs of illness begin to fade. A feeling of well-being is a sure sign that an energetic disturbance in the life force is moving in the direction of balance once again. This is just one among many issues in medicine that are ultimately dependent on one's philosophical orientation.

Another such issue involves one's perspective regarding death. The collective Western denial of death and our attitudes toward material existence have profound and far-reaching influences on the type of health care that we come to expect. Those who accept death as a normal phase that precedes an afterlife are more likely to experience the process as a natural event. Such persons often require less late-stage medical care and fewer interventions compared to those who fight the inevitable as if it is the final end. Physicians, too, will vary in the care that they recommend depending on their beliefs regarding death.

To engage in philosophy is to think deeply on subjects of import in ways that can contribute to the betterment of humankind. Metaphysics contemplates the nature of eternal principles. Conventional science concerns itself with temporal physical laws. Temporal laws do not exist in a vacuum, unaccountable to higher principles, as materialist science would have us believe. Philosophy is not a science and should not be judged on whether or not it progresses in the manner of a science. It is a way of examining worldviews, a way of keeping us honest, a way of holding science accountable.

Science can no longer dissociate itself from philosophy. Philosophy of science first emerged because it became necessary to address the inescapable contradictions inherent in the claims of science. In modern times, those differences have been whitewashed away, thus paving the way for the rise of scientism. Science is not about creating a set of rules that prohibits what we are allowed to investigate, to study, to know. The certainty of science is only temporal. Its laws apply to the material universe and, even then, only in the most provisional of ways. Scientism is just a few short steps away from fascism. Using science as a weapon to suppress free speech stands in direct opposition to scientific principles of free and open inquiry. Fortunately, one of those metaphysical principles that science fails to comprehend seems to be at work here. The eternal Tao is already manifesting as a pendulum swing back in the direction of philosophical concern over scientistic absolutism. Biologist Austin Hughes expresses that concern succinctly:

> The positivist tradition in philosophy gave scientism a strong impetus by denying validity to any area of human knowledge outside of natural science. ... But the last laugh, it seems, remains with the philosophers—for the advocates of scientism reveal conceptual confusions that are obvious upon philosophical reflection. Rather than rendering philosophy obsolete, scientism is setting the stage for its much-needed revival. ...
>
> Continued insistence on the universal competence of science will serve only to undermine the credibility of science as a whole. The ultimate outcome will be an increase of radical skepticism that questions the ability of science to address even the questions legitimately within its sphere of competence. One longs for a new Enlightenment to puncture the pretensions of this latest superstition.[60]

It is my belief that the secularization of knowledge is the fatal flaw of science. To separate the sacred from the profane is an artificial exercise of the dualistic rational mind. All knowledge is sacred. Science is divine. The only way to save medicine from its soulless state of materialistic sterility will be to inject an attitude of respectful receptivity toward all phenomena back into our collective medical belief system. Here, philosopher Manly Hall describes the stakes involved in metaphysics, the missing ingredient in our contemporary approach to the art and science of healing:

> We might say that metaphysics is a vast structure, a noble temple, with its footings in the foundation of the universe, and the vast arch of heaven itself its only roof. The ages have sought for truth. Hundreds of millions have lived to achieve it, and millions have died for it. Heroes, martyrs, sages, saints and prophets, world saviors and demigods of forgotten ages, are the priests of this great house. The gates of this sanctuary are to be approached only with reverence.[61]

We should certainly add scientists to Hall's list of those who have "sought for truth." However, while the accumulation of all the scientific knowledge ever produced may begin to describe creation, it does so only up to a point. It barely scratches the surface of consciousness and gives us only a glimmer of divine principles. Although one may conscientiously pay heed to all physical laws regarding human health, health can nevertheless remain a tenuous thing when higher spiritual principles go neglected.

It is possible to bring medicine to life once again with the assistance of open discussion and debate. Freethinking is always preferable to no thinking at all. Freedom of thought is paramount; it should and will always supersede the scientific facts, regardless of their degree of certainty. Factual certainty will never eliminate the need for ongoing receptivity to new ideas and the potential future implications of unexplained phenomena, all of which may someday render that which was previously thought certain, debatable.

We do not experience life through science alone. Science is but one filter through which to view existence. Philosophy is another. While it has its scientific dimension, life is also art, myth, metaphor, metaphysics, love and, above all else, personal and intimate experience. The many dimensions of human experience must be allowed to have their say, to make their rightful contributions to the art and science of healing. And medical science must acknowledge the legitimate value of those contributions.

Life, death, suffering, and illness will always remain mysteries to some extent. Science is a valuable tool for material living but it is just one small man-made part of that greater mystery. Professional athletes die from heart attacks, chain smokers live to be eighty, and some young children sadly succumb to brain tumors. The best that we can do is to strive to restore health and provide comfort within the limits of our knowledge and abilities. There are never any guarantees. Authentic healing is a science and art that is informed by metaphysical perspective and philosophical beliefs. Patients and healers alike must be allowed once again to think freely and to debate openly matters of import that have a bearing on our approach to illness and our methods of healing the sick.

Glossary Of Terms

The following is a list of terms defined especially as they pertain to the spirit and intent with which they are used in this book. As such, these definitions may differ from conventional and academic definitions and standards.

Abstraction – considering an actual event or thing in abstract form. Removing the specific identifying features of an object or experience in order to yield a theory or idea of an abstract nature. Medical abstraction results in patients that are identified by their physical exam findings, lab values, and imaging tests and not by individual traits or the personal circumstances of their illnesses.

Anecdote – a derogatory and dismissive term that minimizes the value of personal accounts of experiential knowledge. Conventional medicine uses the term to express its view that subjective information is untrustworthy. The common usage of this term is indicative of the extent to which distrust of firsthand experience has grown.

Conformism – (also *universalism*) a conventional medical belief that disease states, diagnostic criteria, treatment interventions, and research results must be uniform in order to be relevant and of value. This theory is at the root of cookie-cutter medicine, which seeks to apply universal standards to large populations of patients. It flies in the face of holistic reality, which demonstrates that patients and their illnesses are not homogeneous; they are unique and highly dependent upon individual circumstance.

Consciousness – used in this book, it means the sum total of all mental, emotional, and spiritual experience. The psychic universe as a whole. The subjective dimension of individual experience. In common usage it refers to psychological self-awareness.

Conventional science – science as conceived and practiced according to the *–isms* defined in this book. Science based on metaphysical assumptions consistent with materialism, reductionism, mechanism, rationalism, objectivism, empiricism, conformism, and dualism.

Dualism – a conceptualization of the rational mind that divides existence into two contrasting halves. Mind and matter, good and evil, and, in medicine, mind and body are examples of dualism. In contrast to holism, which presupposes all of existence as one continuous whole.

Emergentism - (see *epiphenomenalism*) according to this theory, consciousness is an emergent property of matter. Consciousness somehow emerges from matter once matter achieves a sufficient level of complexity to generate mental phenomena. More than the sum of its material origins, consciousness is nevertheless dependent on matter for its existence. This, like epiphenomenalism, is just another materialist theory with no basis in actual science.

Empiricism – a theory of knowledge that relies on information gained through observation and experiment using the senses. Conventional empiricism defines sensory experience in objective terms, in terms of the anatomical and functional nature of human sensory organs. In contrast to *radical empiricism*, which does not exclude information provided by the subjective side of sensory experience and, more importantly, does not exclude the contents of consciousness.

Epiphenomenalism – the theory that all mental phenomena are caused by physical events in the brain. It also posits that mental events are not capable of influencing material reality, thus rendering psychosomatic relationships impossible. Thoughts, emotions, memories, etc, are just epiphenomena, byproducts of brain electrochemistry. Mischaracterized as a mind-body philosophy because it is, in reality, a materialist philosophy in disguise.

Epistemology – the study of knowledge. A branch of philosophy concerned with the nature of knowledge, how we know things, why we know things, and how we know the things that we know are true.

Green medicine – medicine conceived in its broadest, most inclusive, most holistic terms. It recognizes all perspectives and all healing modalities as having potential value. It recognizes the impact of body, heart, mind, soul, family, community, environment, and ecosystem upon human health.

Holism – a theory that views the whole as greater than the sum of its parts. All phenomena are interconnected and cannot be adequately understood in isolation from the whole. A superficial understanding of holism in medicine is taken to mean that patients' psychological needs must be tended to in addition to their physical needs. This is not the same as saying that mind, body, heart, and soul are a seamless whole. Any intervention on one particular level has repercussions for all levels since those levels are really just artificial categories created by left-brain thought.

Holistic reality – reality inclusive of all material and immaterial phenomena. Objective and subjective reality combined. Reality as it is, before it has been processed through the prism of the rational mind. Reality as constructed by science is a poor model of holistic reality because it excludes many aspects of the everyday experience of life.

Left-brain mode – used in this book to refer to a mode of thought associated with the conventional scientific worldview. It is rational, analytical, logical, quantitative, linear, dualistic, black and white, and either/or in its approach to life and living. Although it is a natural way of thinking for some, it creates problems for science when used myopically without right-brain input.

Mechanism – a belief that all natural phenomena must be able to be explained in physical terms, according to material principles of cause and effect. As such, it does not account for mind-body interactions. In medicine, it leads to the overly simplistic perception of the human body as a complex machine that can be restored to health by fixing or replacing malfunctioning or broken parts.

Metaphysics – a philosophy that encompasses all of existence and concerns itself with first principles. Metaphysics is inclusive of science, religion, cosmology, ontology, epistemology, and much more.

Monism – a philosophy that assumes all things derive from one original source. Most monist theories are materialist theories, in which all things are thought to be material in nature. One notable exception is *idealism*, which states that all phenomena derive from, or are dependent on, mind or consciousness. Although material monist theories are attempts to resolve the dilemmas of dualism, they derive from a left-brain perspective, which is the very source of dualism itself.

Objectivism – objectivity is a defining principle of science and medicine. The various layers of meaning of objectivism all have a common intent, which is to diminish the significance and value of firsthand subjective experience. Thought by most to be a fundamental feature of science, objectivism is, in reality, a left-brain value that has been imposed upon science.

Ontology – the study of the nature of being, of existence, of reality. Ontology is relevant to this book because how we define reality has practical implications for how we practice medicine. The metaphysics of conventional medicine defines subjective experience as either non-existent or not relevant, whereas holism assumes that subjective experience is a fundamental feature of reality if not equivalent to reality itself.

Positivism – a philosophical theory grounded in logic and rationalism that rejects all metaphysical and religious concerns. Although now discredited, it was the dominant school of thought in academia for a long time. It is strongly associated with the beliefs of scientism and skepticism.

Proximate causation – assignment of cause and effect according to mechanistic and reductionistic principles. Phenomena that occur close together in space and/or time are commonly mistaken to be causally related when they are, in fact, related only by association, without having a cause and effect connection. Most medical phenomena are explained in terms of proximate causation, while other phenomena that point to broader, deeper forms of holistic causation are often ignored.

Pseudoscience – (also *junk science*) a derogatory misnomer applied to scientific methods and systems of inquiry that do not subscribe to the *–isms* of conventional science. A term used to express prejudice in favor of conventional scientific thinking.

Radical empiricism (see Empiricism)

Rationalism – a belief that rational thought is the most dependable route to certain knowledge. Left-brain, objective, logical, quantitative, reasoned analysis is assumed to be superior to the information provided by firsthand, personal, subjective experience. In medicine, rationalist influences lead to the formulation of a multitude of theories regarding health and illness that lead to poor patient outcomes, primarily because they fail the practical test of holistic reality.

Reductionism – the tendency to explain the nature and function of whole entities in terms of their parts. Complex systems like the human body are treated as if solutions to problems can be found in their parts. Reductionist influences cause medicine to seek answers to health problems in ever-smaller parts such as cells, chromosomes, and biochemical constituents. In this sense, reductionism moves in a direction opposite to holism.

Right-brain mode – refers to a mode of thought that is qualitative, nonlinear, looks for patterns, is capable of seeing the big picture, is comfortable with ambiguity, makes use of intuition and hunches, values feelings, and can think in metaphors. Although right-brain thought is more aligned with holism, genuine holism is inclusive of both right and left-brain qualities. Science in general is biased in favor of left-brain factors and neglects valuable right-brain input.

Scientism – faith in science as the one and only true means of obtaining useful knowledge of reality. Typically associated with materialism and atheism, scientism sees its role as the defender of conventional science in opposition to all forms of religious belief, spirituality, New Age thought, and so-called pseudoscience.

Skeptic – a term that devotees of scientism use to refer to themselves. Skeptics do not subscribe to the *scientism* label because they think that their ideas are based in science, not faith. Skeptics think of themselves as open-minded when, in fact, their default belief is that all things are false until proven true in accordance with the conventional *–isms* of science.

Subjectivity – all contents of consciousness are subjective. All that we feel, think, dream, and experience falls under the category of subjectivity. Subjective knowledge is believed by science and medicine to be inferior to objective information. It is the dimension that science mistakenly believes can be excluded from its deliberations and decision-making processes.

Synchronicity – a term coined by Carl Jung that refers to phenomena that appear on the surface to be related but which are found to have no discernable causal connection. Conventional science ignores such "meaningful coincidences" even though they may be related at deeper levels of holistic reality not recognized or understood by science.

Ultimate causation – (also *final causation*) a theoretical ideal that holism strives to achieve. While it may be impossible to know true ultimate causation, it is possible to seek deeper, more comprehensive solutions to health problems. The closer medical interventions come to resolving final causes, the more effective and long lasting the outcomes tend to be. Because conventional medical treatment focuses on local, proximate causes, its results are usually short lived and can be counterproductive.

Vitalism – a medical theory that posits the existence of an animating life force that transcends chemical, biological, and physical forces, and without which the human body would be a lifeless bag of protoplasm. Known as *chi* in Chinese medicine and *prana* in Ayurvedic medicine, the *vital force* is more than just a bioenergetic phenomenon. It constitutes an innate wisdom endowed by nature that maintains homeostasis and restores health after illness. The vital force makes all instances of healing and self-healing possible.

Epigraph Credits & Permissions

Introduction
Rupert Sheldrake, *The Science Delusion: Freeing the Spirit of Enquiry* (Later published as *Science Set Free*) Coronet, Hodder & Stoughton Ltd, London, 2012, p. 328.

Chapter 1
B. Alan Wallace, *The Taboo of Subjectivity: Toward a New Science of Consciousness*, Oxford University Press, 2000. P. 186.

Chapter 2
C. G. Jung, *The Collected Works*, Princeton University Press, Princeton, NJ, 1971-1986, 13: par 229.

Quote attributed to Nikola Tesla

Chapter 3
Ken Wilber, *No Boundary: Eastern and Western Approaches to Personal Growth*, Shambhala; Reprint edition, 2001

Chapter 4
Rupert Sheldrake, *The Science Delusion: Freeing the Spirit of Enquiry* (Later published as *Science Set Free*), Coronet, Hodder & Stoughton Ltd, London, 2012, p. 30.

Rupert Sheldrake, *The Science Delusion: Freeing the Spirit of Enquiry* (Later published as *Science Set Free*) Coronet, Hodder & Stoughton Ltd, London, 2012, p. 165.

Chapter 5
Gai Eaton, *Science and the Myth of Progress*, Edited by Mehrdad M. Zarandi, World Wisdom, Inc., Bloomington, IN, 2003, p. 201.

Words by Robert Hunter; music by Bob Weir, "Playing in the Band," © Ice Nine Publishing Company. Used with permission

Chapter 6

Werner Heisenberg, *The Physicist's Conception of Nature*, Harcourt, Brace, 1958, p. 24.

Chapter 7

Paul Feyerabend, *The Tyranny of Science*, Polity Press, Malden, MA, 2011, p. 40.

Don Salmon and Jan Maslow, *Yoga Psychology and the Transformation of Consciousness*, Paragon House, St. Paul, MN, 2007, p. 231.

Chapter 8

Thomas Szasz, *The Untamed Tongue: A Dissenting Dictionary*, Open Court Publishing Company, 1990.

Chapter 9

Black Elk, *The Sacred Pipe: Black Elk's Account of the Seven Rites of the Oglala Sioux*, 1953.

Alan Watts. *Tao: The Watercourse Way*. New York: Pantheon, 1975.

Chapter 10

Gai Eaton, *Science and the Myth of Progress*, Edited by Mehrdad M. Zarandi, World Wisdom, Inc., Bloomington, IN, 2003, p. 195.

T. S. Eliot, *The Rock*, London: Faber & Faber, 1934.

Chapter 11

Alfred North Whitehead, *Modes of Thought*, Macmillan, New York, 1938, p. 189.

Chapter 12

Scientists tell us their favourite jokes, The Observer, Dec 28, 2013

http://www.theguardian.com/science/2013/dec/29/scientists-favourite-jokes

Chapter 13

Words by Robert Hunter; music by Jerry Garcia, "Terrapin Station," © Ice Nine Publishing Company. Used with permission

Iain McGilchrist, *The Master and his Emissary: The Divided Brain and the*

Making of the Western World, Yale University Press, New Haven and London, 2009, p. 460.

B. Alan Wallace, *The Taboo of Subjectivity: Toward a New Science of Consciousness*, Oxford University Press, 2000, p. 128.

Chapter 14

Isaac Bonewits, *The Impact of Scientism on Competing Faiths*, 2005.

http://www.neopagan.net/Scientism.html

B. Alan Wallace, *The Taboo of Subjectivity: Toward a New Science of Consciousness*, Oxford University Press, 2000. P. 186.

Chapter 15

Huston Smith, *Science and the Myth of Progress*, Edited by Mehrdad M. Zarandi, World Wisdom, Inc., Bloomington, IN, 2003, p. 234.

Max Plank, *The Observer*, London, January, 25, 1931.

Chapter 16

J.T. Kent, MD, *Lectures on Homeopathic Philosophy*, Examiner Printing House, Lancaster, PA, 1900.

Chapter 17

Manly P. Hall, *Words to the Wise: A Practical Guide to the Esoteric Sciences*, The Philosophical Research Society, Inc., Los Angeles, CA, 1964 (2009 Edition) p. 61.

Physics Legend Freeman Dyson on the One Thing We Just Don't Get About Science, David Freeman, Huffington Post, May 6, 2014.

Manly P. Hall, *Words to the Wise: A Practical Guide to the Esoteric Sciences*, The Philosophical Research Society, Inc., Los Angeles, CA, 1964 (2009 Edition) p. 119.

Notes

Chapter 2

1. Richard Holmes, *The Age of Wonder: How the Romantic Generation Discovered the Beauty and Terror of Science*. London: Harper Press. 2008, p. 449.

2. Thomas Merton, *Conjectures of a Guilty Bystander*, New York, Doubleday, 1966, p. 230.

3. Rene Guenon, *The Crisis of the Modern World*, Sophia Perennis, Hillsdale, NY. 2001 (original publication 1942), pp. 53-54.

Chapter 3

4. Mircea Eliade, *The Two and the One*, Harwell Press, pp. 156-57.

Chapter 5

5. Rene Descartes, *Principles of Philosophy*, 1641, Part 1, Article 7.

6. Prasad V, Vandross A, Toomey C, et al. "A Decade of Reversal: An Analysis of 146 Contradicted Medical Practices." *Mayo Clin Proceedings*. 2013; 88(8): 790–798.

7. Paul Feyerabend, *The Tyranny of Science*, Polity Press, Malden, MA, 2011, p. xi.

Chapter 6

8. Don Salmon and Jan Maslow, *Yoga Psychology and the Transformation of Consciousness*, Paragon House, St. Paul, MN, 2007, pp. 29-30.

9. B. Alan Wallace, *The Taboo of Subjectivity: Toward a New Science of Consciousness*, Oxford University Press, 2000, p. 88.

10. B. Alan Wallace, *The Taboo of Subjectivity: Toward a New Science of Consciousness*, Oxford University Press, 2000, p. 161.

11. B. Alan Wallace, *The Taboo of Subjectivity: Toward a New Science of Consciousness*, Oxford University Press, 2000, p. 29.

Chapter 7

12. Samuel Hahnemann, MD, *Organon of the Medical Art*, edited and annotated by Wenda Brewster O'Reilly, adapted from the sixth edition of the Organon, 1842, Birdcage Books, Redmond, WA, 1996, p. 60.

13. *Oxford Dictionary*.

14. *The Practitioner's Medical Dictionary*, Gould, 1917.

15. *Taber's Cyclopedic Medical Dictionary*, 2001.

16. *Stedman's Medical Dictionary*, 1982.

17. Harris L. Coulter, *Divided Legacy: A History of the Schism in Medical Thought, Volume I: The Patterns Emerge, Hippocrates to Paracelsus*, Center for Empirical Medicine, 1975, Second Printing, 1994, p. 497.

18. Harris L. Coulter, *Divided Legacy: A History of the Schism in Medical Thought, Volume I: The Patterns Emerge, Hippocrates to Paracelsus*, Center for Empirical Medicine, 1975, Second Printing, 1994, p. 502.

19. Harris L. Coulter, *Divided Legacy: A History of the Schism in Medical Thought, Volume I: The Patterns Emerge, Hippocrates to Paracelsus*, Center for Empirical Medicine, 1975, Second Printing, 1994, p. 506.

20. Harris L. Coulter, *Divided Legacy: A History of the Schism in Medical Thought, Volume I: The Patterns Emerge, Hippocrates to Paracelsus*, Center for Empirical Medicine, 1975, Second Printing, 1994, pp. 500-501.

21. B. Alan Wallace, *The Taboo of Subjectivity: Toward a New Science of Consciousness*, Oxford University Press, 2000, p. 59.

22. B. Alan Wallace and Brian Hodel, *Embracing Mind: The Common Ground of Science & Spirituality*, Shambhala, Boston, MA, 2008, p. 74.

Chapter 9

23. Joseph Levine, "Materialism and qualia: the explanatory gap," *Pacific Philosophical Quarterly*, 1983. 64: 354-361.

24. David Chalmers, "Facing Up to the Problem of Consciousness," *Journal of Consciousness Studies*, 2 (3), 1995.

Chapter 10

25. *Taber's Cyclopedic Medical Dictionary*, 2001.

Chapter 12

26. Stolberg HO, Norman G, Trop I (2004). "Randomized controlled trials". Am J Roentgenol 183 (6): 1539–44.

27. Meldrum ML. "A brief history of the randomised controlled trial. From oranges and lemons to the gold standard." *Hematol Oncol Clin North Am* 2000;14 (4):745–60.

28. Marcia Angell, "Drug Companies & Doctors: A Story of Corruption," *The New York Review of Books*, Jan 15, 2009. http://www.nybooks. com/articles/archives/2009/jan/15/drug-companies-doctorsa-story-of-corruption/

29. *Persuading the Prescribers: Pharmaceutical Industry Marketing and its Influence on Physicians and Patients*, The Pew Charitable Trusts, Nov 11, 2013,
http://www.pewhealth.org/other-resource/persuading-the-prescribers-pharmaceutical-industry-marketing-and-its-influence-on-physicians-and-patients-85899439814

30. "What conclusions has Clinical Evidence drawn about what works, what doesn't based on randomised controlled trial evidence?," ClinicalEvidence, 2014.
http://clinicalevidence.bmj.com/x/set/static/cms/efficacy-categorisations.html

31. Ioannidis, JPA, "Why Most Published Research Findings Are False." PLoS Med 2(8): e124. doi:10.1371/journal.pmed.0020124, 2005.

32. Ioannidis, JPA, "Why Most Published Research Findings Are False." PLoS Med 2(8): e124. doi:10.1371/journal.pmed.0020124, 2005.

33. A Decade of Reversal: "An Analysis of 146 Contradicted Medical Practices," *Mayo Clinic Proceedings*, Volume 88, Issue 8, pp. 790-798, August 2013.

34. Mark Hyman, MD, "Science for Sale: Protect Yourself From Medical Research Deception," Huffington Post, Oct 2010. http://

www.huffingtonpost.com/dr-mark-hyman/dangerous-spin-doctors-7-_b_747325.html

Chapter 13

35. Rupert Sheldrake, *The Science Delusion: Freeing the Spirit of Enquiry,* (Later published as *Science Set Free*) Coronet, Hodder & Stoughton Ltd, London, 2012, p. 292.

36. Manly P. Hall, *Words to the Wise: A Practical Guide to the Esoteric Sciences,* The Philosophical Research Society, Inc., Los Angeles, CA, p. 121.

37. Rene Guenon, *The Crisis of the Modern World*, Sophia Perennis, Hillsdale, NY. 2001 (original publication 1942), p. 45.

Chapter 14

38. Leon Lederman and Dick Teresi, *God Particle: If the Universe Is the Answer, What Is the Question?*, reprint of the original 1993 version by First Mariner Books, 2006, p. 22.

39. Leon Lederman and Dick Teresi, *God Particle: If the Universe Is the Answer, What Is the Question?*, reprint of the original 1993 version by First Mariner Books, 2006, p. 24.

40. For a nice overview of the history of the Skeptic movement: "Paul Kurtz, founder of modern Skepticism, dies at 86," Oct 2012, philosophyforlife.org

41. Carl Sagan, "Wonder and Skepticism," Skeptical Inquirer, Volume 19.1, Jan 1995. *http://www.csicop.org/si/show/wonder_and_skepticism*

42. Carl Sagan, *The Demon-Haunted World: Science as a Candle in the Dark,* Ballantine Books, 1997, p. 204.

43. Quote attributed to Neil deGrasse Tyson, Director Hayden Planetarium.

44. "A letter to the TEDx community on TEDx and bad science," October 2013. http://blog.tedx.com/post/37405280671/a-letter-to-the-tedx-community-on-tedx-and-bad-science

45. Nelson W. Polsby PS, Vol. 17, No. 4. (Autumn, 1984), pp. 778-

781. Pg. > 779. http://childhealthsafety.wordpress.com/2012/11/20/chs-medical-myth-smasher-1-the-plural-of-anecdote-is-data-is-the-correct-original-quotation/

46. "Plausibility bias and the controversy around homeopathy," Homeopathy Research Institute, Newsletter, Issue 13, Summer 2011. http://www.homeoinst.org/sites/default/files/newsletters/HRI_Newsletter13_Summer2011.pdf

47. Stephen Bond, "Why I Am No Longer a Skeptic."

48. The Pathology of Organized Skepticism, *Journal of Scientific Exploration*, Volume 16, No. 1, pp. 125-28, 2002. http://www.scientificexploration.org/journal/jse_16_1_leiter.pdf

49. New Atheism, Wikipedia, *http://en.wikipedia.org/wiki/New_atheism*

50. PZ Myers, "High Priest Epstein in Newsweek," ScienceBlogs, June 2007. http://scienceblogs.com/pharyngula/2007/06/14/high-priest-epstein-in-newswee/

Chapter 15

51. B. Alan Wallace and Brian Hodel, *Embracing Mind: The Common Ground of Science & Spirituality*, Shambhala, Boston, MA, 2008, pp. 142-143.

52. B. Alan Wallace and Brian Hodel, *Embracing Mind: The Common Ground of Science & Spirituality*, Shambhala, Boston, MA, 2008, p. 145.

53. William James, *Essays in Radical Empiricism*, Harvard University Press, 1976, p. 22.

54. B. Alan Wallace, *The Taboo of Subjectivity: Toward a New Science of Consciousness*, Oxford University Press, 2000, p. 75.

55. B. Alan Wallace and Brian Hodel, *Embracing Mind: The Common Ground of Science & Spirituality*, Shambhala, Boston, MA, 2008, p. 201.

56. Ken Wilber, *The Spectrum of Consciousness*, Quest Books, Wheaton, IL., 1993 (original 1977), p. 24-25.

57. Ken Wilber, *The Spectrum of Consciousness*, Quest Books, Wheaton, IL., 1993 (original 1977), p. 28.

58. Ken Wilber, *The Spectrum of Consciousness*, Quest Books, Wheaton, IL., 1993 (original 1977), p. 66.

Chapter 17

59. Rupert Sheldrake, *The Science Delusion: Freeing the Spirit of Enquiry*, (Later published as *Science Set Free*) Coronet, Hodder & Stoughton Ltd, London, 2012, p. 328.

60. Austin L. Hughes, "The Folly of Scientism," *The New Atlantis*, Number 37, Fall 2012.

61. Manly P. Hall, *Words to the Wise: A Practical Guide to the Esoteric Sciences*, The Philosophical Research Society, Inc., Los Angeles, CA, p. 146.

Bibliography

Angell, MD, Marcia, "Drug Companies & Doctors: A Story of Corruption," *The New York Review of Books*, Jan 15, 2009.

Appleyard, Bryan, *Understanding the Present: Science and the Soul of Modern Man*, Doubleday, New York, 1993.

Black Elk, *The Sacred Pipe: Black Elk's Account of the Seven Rites of the Oglala Sioux*, 1953.

Bond, Stephen, "Why I Am No Longer a Skeptic."

Bonewits, Isaac, *The Impact of Scientism on Competing Faiths*, 2005.

Burtt, E. A., *The Metaphysical Foundations of Modern Science*, Dover Publications, Inc, New York, 2003.

Chalmers, David, "Facing Up to the Problem of Consciousness," *Journal of Consciousness Studies*, 2 (3), 1995.

Coulter, Harris L., *Divided Legacy: A History of the Schism in Medical Thought, Volume I: The Patterns Emerge, Hippocrates to Paracelsus*, Center for Empirical Medicine, 1975, Second Printing, 1994.

"A Decade of Reversal: An Analysis of 146 Contradicted Medical Practices," *Mayo Clinic Proceedings*, Volume 88, Issue 8, pp. 790-798, August 2013.

Descartes, Rene, *Principles of Philosophy*, 1641.

Eliade, Mircea, *The Two and the One*, Harwell Press, 1962.

Feyerabend, Paul, *The Tyranny of Science*, Polity Press, Malden, MA, 2011.

Godfrey-Smith, Peter, *Theory and Reality: an introduction to the philosophy of science*, The University of Chicago Press, 2003.

Guenon, Rene, *The Crisis of the Modern World*, Sophia Perennis, Hillsdale, NY., (original publication 1942) 2001.

Hahnemann, MD, Samuel, *Organon of the Medical Art*, edited and annotated by Wenda Brewster O'Reilly, adapted from the sixth edition of the Organon, 1842, Birdcage Books, Redmond, WA, 1996.

Hall, Manly P., *Words to the Wise: A Practical Guide to the Esoteric Sciences*, The Philosophical Research Society, Inc., Los Angeles, CA, (original publication 1964) 2009.

Heisenberg, Werner, *The Physicist's Conception of Nature*, Harcourt, Brace, 1958.

Holmes, Richard, *The Age of Wonder: How the Romantic Generation Discovered the Beauty and Terror of Science*. London: Harper Press. 2008.

Hughes, Austin L., "The Folly of Scientism," *The New Atlantis*, Number 37, Fall 2012.

Huxley, Aldous, *The Perennial Philosophy*, Harper Perennial Modern Classics, 2009.

Hyman, MD, Mark, "Science for Sale: Protect Yourself From Medical Research Deception," Huffington Post, Oct 2010.

Inner Knowing: Consciousness, Creativity, Insight, and Intuition, Edited by Helen Palmer, Jeremy P. Tarcher/Putnam, New York, 1998.

Ioannidis, JPA, "Why Most Published Research Findings Are False." PLoS Med 2(8): e124. doi:10.1371/journal.pmed.0020124, 2005.

James, William, *Essays in Radical Empiricism*, Harvard University Press, 1976.

Jung, C. G., *The Collected Works*, Princeton University Press, Princeton, NJ, 1971-1986.

Lederman, Leon and Teresi, Dick, *God Particle: If the Universe Is the Answer, What Is the Question?*, reprint of the original 1993 version by First Mariner Books, 2006.

"A letter to the TEDx community on TEDx and bad science," October 2013.
http://blog.tedx.com/post/37405280671/a-letter-to-the-tedx-community-on-tedx-and-bad-science

Levine, Joseph, "Materialism and qualia: the explanatory gap," *Pacific Philosophical Quarterly*, 1983. 64: 354-361.

McGilchrist, Iain, *The Master and his Emissary: The Divided Brain and the Making of the Western World*, Yale University Press, New Haven and London, 2009.

Meldrum ML. "A brief history of the randomised controlled trial. From oranges and lemons to the gold standard." *Hematol Oncol Clin North Am* 2000;14 (4):745–60.

Merchant, Carolyn, *The Death of Nature: Women, Ecology and the Scientific Revolution*, HarperOne, 1989.

Merton, Thomas, *Conjectures of a Guilty Bystander*, New York, Doubleday, 1966.

Milton, Richard, *Alternative Science: Challenging the Myths of the Scientific Establishment*, Park Street Press, Rochester, VT, 1996.

Myers, PZ, "High Priest Epstein in Newsweek," ScienceBlogs, June 2007.

New Atheism, Wikipedia.org

"The Pathology of Organized Skepticism," *Journal of Scientific Exploration*, Volume 16, No. 1, pp. 125-28, 2002.

"Paul Kurtz, founder of modern Skepticism, dies at 86," Oct 2012, philosophyforlife.org.

Persuading the Prescribers: Pharmaceutical Industry Marketing and its Influence on Physicians and Patients, The Pew Charitable Trusts, Nov 11, 2013.

"Plausibility bias and the controversy around homeopathy," Homeopathy Research Institute, Newsletter, Issue 13, Summer 2011.

"The Plural of Anecdote is Data" – Is The Correct Original Quotation,

Child Health Safety, Nov 2012. http://childhealthsafety.wordpress.com/

The Practitioner's Medical Dictionary, Gould, 1917.

Prasad V, Vandross A, Toomey C, et al. "A Decade of Reversal: An Analysis of 146 Contradicted Medical Practices." *Mayo Clin Proceedings*. 2013; 88(8): 790–798.

Sagan, Carl, *The Demon-Haunted World: Science as a Candle in the Dark*, Ballantine Books, 1997.

Sagan, Carl, "Wonder and Skepticism," Skeptical Inquirer, Volume 19.1, Jan 1995.

Salmon, Don and Maslow, Jan, *Yoga Psychology and the Transformation of Consciousness*, Paragon House, St. Paul, MN, 2007.

Science and the Myth of Progress, Edited by Mehrdad M. Zarandi, World Wisdom, Inc., Bloomington, IN., 2003.

Sheldrake, Rupert, *The Science Delusion: Freeing the Spirit of Enquiry* (Later published as *Science Set Free*) Coronet, Hodder & Stoughton Ltd, London, 2012.

Smith, Huston, *Beyond the Postmodern Mind: The Place of Meaning in a Global Civilization*, Quest Books, Wheaton, IL., 2003.

Sorell, Tom, *Scientism: Philosophy and the Infatuation with Science*, Routledge, London and New York, 1994.

Stedman's Medical Dictionary, 1982.

Stolberg HO, Norman G, Trop I (2004). "Randomized controlled trials". Am J Roentgenol 183 (6): 1539–44.

Szasz, Thomas, *The Untamed Tongue: A Dissenting Dictionary*, Open Court Publishing Company, 1990.

Taber's Cyclopedic Medical Dictionary, 2001.

Wallace, B. Alan and Hodel, Brian, *Embracing Mind: The Common Ground of Science & Spirituality*, Shambhala, Boston, MA, 2008.

Wallace, B. Alan, *The Taboo of Subjectivity: Toward a New Science of Consciousness*, Oxford University Press, 2000.

Watts, Alan, *Tao: The Watercourse Way*. New York: Pantheon, 1975.

"What conclusions has Clinical Evidence drawn about what works, what doesn't based on randomised controlled trial evidence?," ClinicalEvidence, 2014.

Whitehead, Alfred North, *Modes of Thought*, Macmillan, New York, 1938.

Wilber, Ken, *No Boundary: Eastern and Western Approaches to Personal Growth*, Shambhala; Reprint edition, 2001.

Wilber, Ken, *The Spectrum of Consciousness*, Quest Books, Wheaton, IL., (original publication 1977) 1993.

About The Author

Larry Malerba, DO, DHt is a physician, educator, and author whose mission is to build bridges between holistic healing, conventional medicine, and spirituality. His first book, *Green Medicine: Challenging the Assumptions of Conventional Health Care*, is a sweeping overview of holistic theory. He has written articles for the American Holistic Medical Association, *Huffington Post*, *Reality Sandwich*, and *Natural News*.

Dr. Malerba is board certified in Homeotherapeutics, is Clinical Assistant Professor at New York Medical College, past president of the Homeopathic Medical Society of the State of New York, and past vice president of the American Institute of Homeopathy. He is a dual citizen of the U.S. and Ireland and has a private practice in upstate New York.

You can follow DocMalerba on Facebook and Twitter.
You can learn more about Dr. Malerba at his website:
www.SpiritScienceHealing.com

Also By Larry Malerba, DO

"GREEN MEDICINE will change the way you think
about healthcare and healing."

Made in the USA
Middletown, DE
19 November 2014